CITY GREEN: INNOVATIVE GREEN INFRASTRUCTURE SOLUTIONS FOR DOWNTOWNS AND INFILL LOCATIONS

ACKNOWLEDGMENTS

This report was prepared with the assistance of Horsley-Witten Group.

EPA Project Leads: Lisa Hair, Office of Water, and Melissa Kramer, Office of Sustainable Communities

If you have questions about this publication, please contact:

Melissa Kramer
Office of Sustainable Communities
U.S. Environmental Protection Agency
1200 Pennsylvania Avenue NW (MC 1807T)
Washington, DC 20460
Tel 202-564-8497
Kramer.melissa@epa.gov

Reviewers:

Lynn Desautels, Chitra Kumar, and Megan Susman - EPA Office of Sustainable Communities

Katelyn Amraen, Leah Germer (ORISE Fellow), and Jamie Piziali - EPA Office of Water

Trish Garrigan and Rosemary Monahan - EPA Region 1

Rabi Kieber and Maureen Krudner - EPA Region 2

Dominique Lueckenhoff - EPA Region 3

Anne Keller and Christine McKay - EPA Region 4

Robert Newport - EPA Region 5

Suzanna Perea - EPA Region 6

David Doyle - EPA Region 7

Stacey Eriksen - EPA Region 8

Carolyn Mulvihill and Luisa Valiela - EPA Region 9

Jeremy Chadwick, Alan Fody, and Jessica Noon - Philadelphia Water Department

George Grinton - City of Aiken, South Carolina, Engineering and Utilities Department

David Misky - Redevelopment Authority of the City of Milwaukee

Mercy Davison - Town of Normal, Illinois

Mark Doneux and Anna Eleria - Capitol Region Watershed District (Minnesota)

David Pike - City of Santa Fe Public Works Department

Ken MacKenzie - Urban Drainage and Flood Control District (Denver, Colorado)

Tracy Tackett - Seattle Public Utilities

TABLE OF CONTENTS

EXECUTIVE SUMMARY

Communities of all sizes and in all climates are using green infrastructure to manage stormwater where it falls using the natural processes associated with soils and vegetation to capture, slow down, and filter runoff, often allowing it to recharge ground water. Green infrastructure manages stormwater to control flooding from small storms and improve water quality and offers a wide range of other environmental, economic, public health, and social benefits.

This publication is for local governments, private developers, and other stakeholders who help shape redevelopment projects in downtowns and infill locations where development has already occurred. It provides inspiration and helps identify successful strategies and lessons learned for overcoming common barriers to using green infrastructure in these contexts. The examples could encourage cities to adopt policies that would expand the number of projects incorporating similar green infrastructure approaches.

Twelve case studies showcase projects from around the country that have overcome many common challenges to green infrastructure at sites surrounded by existing development and infrastructure. In these cases, space is at a premium, and soil conditions are often unknown or unsuitable for infiltration. The case studies help identify successful strategies and lessons learned for overcoming common problems. The case studies are:

- The Waltham Watch Factory, Waltham, Massachusetts.
- Queens Botanical Garden, Flushing, New York.
- The Kensington Creative and Performing Arts High School, Philadelphia, Pennsylvania.
- The Radian Complex, Philadelphia, Pennsylvania.

- The Sand River Headwaters Green Infrastructure Project, Aiken, South Carolina.
- The Menomonee Valley Industrial Center, Milwaukee, Wisconsin.
- The Uptown Normal Circle, Normal, Illinois.
- The Metro Green Line, St. Paul, Minnesota.
- Stapleton Greenway Park, Denver, Colorado.
- Santa Fe Railyard Park and Plaza, Santa Fe, New Mexico.
- Mint Plaza, San Francisco, California.
- Thornton Creek Water Quality Channel, Seattle, Washington.

Though green infrastructure can be more challenging to implement in redevelopment projects compared to projects in undeveloped areas, the barriers are usually surmountable. These case studies help counter real and perceived obstacles to using green infrastructure in downtowns and infill locations by providing successful examples showing that:

- Careful site planning and selection of practices allow green infrastructure to work on contaminated sites and sites with poor soils.
- Historic properties can incorporate context-sensitive green infrastructure compatible with the historic fabric.
- Green infrastructure fits into highly space-constrained sites.
- Municipalities are removing regulatory obstacles to allow green infrastructure projects.
- Green infrastructure can provide effective stormwater management in arid climates and areas where water rights are a concern.
- Green infrastructure can be a cost-effective approach to stormwater management and can help drive economic development.
- Long-term maintenance can be addressed by thoughtful upfront planning and innovative approaches.

I. INTRODUCTION

Communities of all sizes and in all climates are using green infrastructure to manage stormwater where it falls. Techniques such as permeable pavement, bioswales, rain gardens, and green roofs use the natural processes associated with soils and vegetation to capture, slow down, and filter runoff, often allowing it to recharge ground water.[1] Other techniques like rain barrels and cisterns collect and store water for future use.[2] Green infrastructure manages stormwater to control flooding from small storms and improve water quality. It also offers a wide range of other environmental, economic, public health, and social benefits (Exhibit 1). As developers and local governments recognize the multiple benefits of site-scale green infrastructure, they are increasingly incorporating it into redevelopment projects in downtowns and infill locations.

This publication is for local governments, private developers, and other stakeholders who help shape redevelopment projects in downtowns and infill locations. Twelve case studies showcase projects from around the country that have overcome many common challenges to green infrastructure at sites surrounded by existing development and infrastructure. In these cases, space is at a premium, and soil conditions are often unknown or unsuitable for infiltration. The case studies help identify successful strategies and lessons learned for overcoming common problems. In addition, by documenting the multiple benefits of green infrastructure, particularly in the redevelopment context, these case studies provide inspiration for local governments and private developers who want to use green infrastructure strategies. The case studies could encourage cities to adopt policies that would expand the number of projects

> **Exhibit 1.** Potential benefits of green infrastructure
>
> Green infrastructure can make the most of limited funds by producing multiple benefits with a single investment. These benefits include:
>
> - Improved water quality.
> - Reduced municipal water use.
> - Ground water recharge.
> - Flood risk mitigation for small storms.
> - Increased resilience to climate change impacts such as heavier rainfalls and hotter temperatures.
> - Reduced ground-level ozone.
> - Reduced particulate pollution.
> - Reduced air temperatures in developed areas.
> - Reduced energy use and associated greenhouse gas emissions.
> - Increased or improved wildlife habitat.
> - Improved public health from reduced air pollution and increased physical activity.
> - Increased recreation space.
> - Improved community aesthetics.
> - Cost savings.
> - Green jobs.
> - Increased property values.
>
> For more information about achieving multiple benefits from green infrastructure, see: EPA. *Enhancing Sustainable Communities with Green Infrastructure.* 2014. https://www.epa.gov/smart growth/enhancing-sustainable-communities-green-infrastructure.

incorporating similar green infrastructure approaches. The case studies are:

- The Waltham Watch Factory, Waltham, Massachusetts.

---

[1] For a complete description of different green infrastructure approaches, see: EPA. "What is Green Infrastructure?" https://www.epa.gov/green-infrastructure /what-green-infrastructure. Accessed Apr. 12, 2016.

[2] Green infrastructure also encompasses larger-scale management strategies, including preserving or restoring flood plains, open space, wetlands, and forests. However, this document focuses on site-scale practices.

Exhibit 2. Locations of profiled projects.

- Queens Botanical Garden, Flushing, New York.
- The Kensington Creative and Performing Arts High School, Philadelphia, Pennsylvania.
- The Radian Complex, Philadelphia, Pennsylvania.
- The Sand River Headwaters Green Infrastructure Project, Aiken, South Carolina.
- The Menomonee Valley Industrial Center, Milwaukee, Wisconsin.
- The Uptown Normal Circle, Normal, Illinois.
- The Metro Green Line, St. Paul, Minnesota.
- Stapleton Greenway Park, Denver, Colorado.
- Santa Fe Railyard Park and Plaza, Santa Fe, New Mexico.
- Mint Plaza, San Francisco, California.
- Thornton Creek Water Quality Channel, Seattle, Washington.

Although green infrastructure can be more challenging to implement in redevelopment projects compared to projects in undeveloped areas, the barriers are usually surmountable. These case studies help counter real and perceived obstacles to using green infrastructure in downtowns and infill locations by providing successful examples showing that:

- Careful site planning and selection of practices allow green infrastructure to work on contaminated sites and sites with poor soils.
- Historic properties can incorporate context-sensitive green infrastructure compatible with the historic fabric.
- Green infrastructure works in highly space constrained sites and can even be a better choice than conventional stormwater management approaches.
- Municipalities are removing regulatory obstacles to allow green infrastructure projects.
- Green infrastructure can provide effective stormwater management in arid climates and areas where water rights are a concern.
- Green infrastructure can be a cost-effective approach to stormwater management and can help drive economic development.
- Long-term maintenance can be addressed by thoughtful upfront planning and innovative approaches.

A. CAREFUL SITE PLANNING AND SELECTION OF PRACTICES ALLOW GREEN INFRASTRUCTURE TO WORK ON CONTAMINATED SITES AND SITES WITH POOR SOILS

In many infill locations, developers suspect that soils might be unsuitable for infiltration or that past industrial and commercial activity has polluted the soil and water. However, with early, careful planning to reduce the potential of contaminating ground water with suspected pollutants, green infrastructure can help these sites become attractive assets to the community. Often testing reveals either no or only partial contamination of a site, and site planners can lay out the development to ensure that green infrastructure practices will not mobilize contaminants. Designers can also select practices that function without infiltrating stormwater into the soil, including green roofs and cisterns. In addition, they can cover contaminated soil with an impervious barrier topped with clean soil and vegetation that filter and evapotranspire stormwater before it reaches an underdrain (located above the barrier) that is connected to the stormwater system.

Construction of the Kensington Creative and Performing Arts High School in Philadelphia occurred on a site contaminated by former industrial uses. Designers incorporated green infrastructure by avoiding areas where contaminants might be mobilized, maximizing infiltration in suitable areas, and using techniques that posed no contamination risk, such as underground storage for stormwater and a green roof. Together, these approaches allowed the site to reduce runoff and pollution entering the combined sewer system.

Developers of Mint Plaza in downtown San Francisco learned that the infiltrative capacity of the native soils was much greater than anticipated. The site was designed to manage a 5-year storm event on-site, but actual performance has approached the 25-year storm event,[3] showing that even older, dense downtown locations can be good candidates for green infrastructure and that soil testing early in the design phase can help developers plan green infrastructure that works with the existing conditions.

Designers of the Waltham Watch Factory redevelopment project also faced a site contaminated by past uses. To deal with this challenge, they lined the rain gardens in the interior courtyards with an impermeable geomembrane. The gardens filter runoff from the surrounding roofs and courtyard paving without posing any threat of mobilizing contaminants in the underlying soil.

B. HISTORIC PROPERTIES CAN INCORPORATE CONTEXT-SENSITIVE GREEN INFRASTRUCTURE COMPATIBLE WITH THE HISTORIC FABRIC

In many historic neighborhoods, on-site management of stormwater was once more common than it is today. Semi-permeable gravel or brick pavers, cisterns, open streams, and gardens were a part of the historic fabric. Context-appropriate incorporation of green infrastructure can thus be compatible with historic properties and can even enhance an area's historic character. In the process of helping to restore degraded waterways, green infrastructure can also enhance the historic character of buildings and neighborhoods,

---

[3] A "25-year storm event" is a storm having a 25-year recurrence interval based on historical data. In other words, a storm of that magnitude has a 4 percent chance of happening in any given year.

creating spaces that honor an area's heritage and that residents and visitors cherish.

A mid-19th century watch factory in Waltham, Massachusetts, was redeveloped as a mixed-use development with apartments, offices, and restaurants. It incorporates green infrastructure that reduces polluted runoff, improves wildlife habitat, and connects residents, workers, and visitors to the Charles River, a water body central to the city's heritage.

In Santa Fe, New Mexico, preserving the historic features at an abandoned railyard and incorporating public open space helped garner community support for redevelopment. A 10-acre park that manages stormwater provides a lively community space next to a new arts and entertainment district that includes a reopened historic train depot, a performing arts center, art galleries, restaurants, and shopping.

In Aiken, South Carolina, residents were concerned about alterations to the town's historic parkways. However, once they understood how planned green infrastructure practices would look and function and that they would help protect the nearby Sand River, residents supported the installation.

C.   GREEN INFRASTRUCTURE FITS INTO HIGHLY SPACE-CONSTRAINED SITES

Downtown properties and infill sites are often space limited. To make projects in these areas financially viable, developers have to maximize the developable area. Green infrastructure practices such as green roofs, permeable pavement, underground cisterns, and structural tree planters can work where space is constrained. In some cases, creating shared green infrastructure facilities can allow individual properties to meet stormwater management requirements.

The Radian Complex is a 500-bed student housing and retail center on the northwest edge of the University of Pennsylvania campus in Philadelphia. Before redevelopment, the site was 99 percent impervious and close to other buildings and infrastructure, which limited infiltration opportunities. A green roof, pervious pavers, tree pits and planters, and two underground stormwater detention basins let the project meet stormwater management requirements to reduce runoff volumes by 20 percent while maximizing the area available for retail development.

The Northgate district redevelopment project in Seattle, Washington, incorporates end-of-pipe water quality treatment for a highly impervious 680-acre sub-watershed. The 2.7-acre stormwater facility has become a haven for wildlife and much-needed open space for residents of new senior and multifamily housing, retail customers, and people using the connection between the neighborhood and transit station.

The Stapleton Airport redevelopment is one of the largest infill projects in the country, located just 6 miles from downtown Denver. The developer integrated green infrastructure into parks and open space, creating centralized facilities that meet water quality, flood control, and open space requirements. These areas are now a selling point for the development and a beloved part of the community.

D. MUNICIPALITIES ARE REMOVING REGULATORY OBSTACLES TO ALLOW GREEN INFRASTRUCTURE PROJECTS

In many downtown and infill locations, long-standing regulations can sometimes present barriers to incorporating green infrastructure that discourage developers from pursuing this approach. As more local governments recognize green infrastructure's benefits, they are helping to break down these barriers. Often, a single successful and popular project is enough to change government policy and allow future projects.

The city of St. Paul and the Capitol Region Watershed District incorporated green infrastructure into a new light rail line linking the cities of St. Paul and Minneapolis. They also installed green infrastructure on adjacent streets along the corridor, and the entire area serves as a demonstration project for other developments in the city.

When designers planned Mint Plaza in San Francisco, the city's codes prohibited directing runoff from adjacent roofs to the plaza's infiltration chambers. When the city issued new stormwater design guidelines a few years later, it changed that policy to encourage developers to use green infrastructure such as infiltration chambers to manage runoff on-site, ensuring that future projects will not face this limitation.

The Queens Botanical Garden in New York City was a pilot project for the city Department of Design and Construction's High Performance Buildings Program, which had the goal of making new and renovated public buildings in the city environmentally sustainable. By demonstrating the feasibility of incorporating environmentally sustainable features into new public buildings, the city was able to institute new requirements for future city projects.

E. GREEN INFRASTRUCTURE CAN PROVIDE EFFECTIVE STORMWATER MANAGEMENT IN ARID CLIMATES AND WHERE WATER RIGHTS ARE A CONCERN

Green infrastructure can manage stormwater and conserve water resources in arid regions if designers select plants for their drought tolerance and use other landscaping techniques that reduce the need for irrigation. Even in areas where water rights laws preclude certain practices, green infrastructure can still be an effective approach to stormwater management.

The Santa Fe Railyard Park in New Mexico incorporates shady riparian areas, a dry gulch that fills seasonally with rain, ornamental gardens adapted for dry conditions, and historic

Pueblo gardens into a 10-acre park filled with native and drought-resistant plants. An innovative water-harvesting system compatible with water rights restrictions irrigates the landscape.

The Stapleton Airport redevelopment in Denver uses vegetated swales and constructed wetlands, which satisfy requirements not to retain, reuse, or store runoff. These techniques allow green infrastructure for water quality treatment even in an area where water rights restrictions limit the practices that can be used.

F. GREEN INFRASTRUCTURE CAN BE A COST-EFFECTIVE APPROACH TO STORMWATER MANAGEMENT AND CAN HELP DRIVE ECONOMIC DEVELOPMENT

Project developers are often concerned about costs for green infrastructure, particularly for downtown and infill sites where the range of appropriate practices can be narrower than on sites developed less compactly. However, green infrastructure can be a cost-effective way to meet stormwater requirements in downtowns and infill sites, particularly as new materials are developed and contractors gain experience. In addition, green infrastructure can create projects that appeal to residents and business owners, helping to fill homes and attract businesses. These economic benefits can help municipal governments and private developers justify green infrastructure's cost.

In Normal, Illinois, a $15.5 million redevelopment project to create a new community space in a traffic circle that incorporates innovative stormwater management has led to $160 million in private business investment in the Uptown District. In addition, property values went up 16 percent, and retail sales grew 46 percent. Public education about the multiple environmental, economic, and social benefits helped generate community support for the initial infrastructure investment.

Redevelopment of a former industrial brownfield site into the Menomonee Valley Industrial Center incorporated a centralized green infrastructure stormwater management system that provides the community a new recreational park with access to the Menomonee River. Property values at the site increased 1,400 percent between 2002 and 2009, adding more than $1 million a year to city property tax revenues. The 10 firms at the site had 1,400 employees as of 2015 on what was once an abandoned site.

Developers of San Francisco's Mint Plaza recognized that the plaza would create an outstanding amenity that would pay dividends by making the company's surrounding properties more valuable and desirable. Since the plaza opened, new restaurants, hotels, and cafes have opened nearby, demonstrating the plaza's economic value.

G. LONG-TERM MAINTENANCE CAN BE ADDRESSED BY THOUGHTFUL, UPFRONT PLANNING AND INNOVATIVE APPROACHES

Communities are often concerned about the long-term maintenance of green infrastructure because it requires practices, resources, and expertise different from those already in place for conventional stormwater infrastructure. Indeed, proper maintenance of green infrastructure is critical to its long-term success. Maintenance will be needed to retain the planned water quality benefits and maintain community support, which can diminish if green infrastructure starts to look unkept. Planning in advance for maintenance requirements helps local governments and developers ensure that the benefits of green infrastructure will continue for years to come.

The Menomonee Valley Industrial Center redevelopment in Milwaukee, Wisconsin, incorporates a stormwater park, creating a public amenity that has generated community support for the project and contributed to its overall success. Volunteers planted trees and collected trash during and after construction, which helped generate a sense of community ownership in the project. Volunteers from local schools, businesses, and neighborhood associations continue to regularly plant new trees and shrubs, remove invasive species, and pick up trash.

Normal, Illinois, has a stormwater utility fee that generates dedicated revenue for green infrastructure projects. These funds implement more green infrastructure in the community and ensure long-term maintenance of those projects.

Developers of Mint Plaza in San Francisco formed an independent, nonprofit organization to manage maintenance and programming on the plaza, including a farmers market and arts performances. The organization also hosts private, revenue-generating events at the site to pay expenses. Identifying funds for long-term maintenance at the time of project planning was an important part of the public permitting process because it eased the city's concern about who would be responsible for these costs.

H.   CONCLUSION

These 12 case studies illustrate a range of circumstances in downtown and infill locations where green infrastructure practices perform well. As local governments and developers look for ways to efficiently use development funds, these examples help illustrate that green infrastructure can effectively control stormwater while helping to achieve other environmental, economic, public health, and social goals.

II. WALTHAM WATCH FACTORY
WALTHAM, MASSACHUSETTS

Redevelopment of a historic mill complex along the Charles River improves water quality and the developer's bottom line while providing new public access to the river.

Project type:	Mixed-use development; historic preservation, brownfield redevelopment
Green infrastructure practices:	Pervious pavement, infiltration trenches, rain gardens, and tree plantings
Completion dates:	Phase 1: May 2009 Phase 2: March 2012 Phase 3: 2014

Green infrastructure is a prominent feature in the redeveloped Waltham Watch Factory, a 150-year-old, 12-acre former mill complex in Waltham, Massachusetts, along the Charles River. The three-phase project involved rehabilitating and converting the historic mill complex to a mix of office, residential, retail, and restaurant space. The developer, Watch City Ventures, recognized that stormwater pollution threatened the health of the project's main asset—the Charles River. Watch City Ventures engaged the Charles River Watershed Association to help incorporate green infrastructure into the site design. Pervious pavement, infiltration trenches, and rain gardens now help cleanse stormwater before it enters the Charles River. Building tenants and visitors can use walking paths to reach the river and can enjoy lush gardens in building courtyards, increasing the Watch Factory's appeal and making it more valuable for its owners.

A.   SITE CONTEXT

The project site is on the banks of the Charles River in the city of Waltham, a western suburb of Boston. The Upper/Middle Charles River is 70 miles long, ending at the Watertown Dam where it connects to the Lower Charles River. The Upper/Middle Charles River watershed covers

268 square miles, encompassing all or parts of 32 communities[4] and is part of the most densely populated watershed in New England.[5]

Excessive algae and aquatic plants significantly impair water quality in the Upper/Middle Charles River. Massachusetts was required to develop a Total Maximum Daily Load (TMDL), which sets the maximum amount of a single pollutant that can enter a waterway while still allowing it to meet water quality standards. The TMDL determined that phosphorus loads must be reduced by 50 percent through a 66 percent reduction from wastewater treatment plants and a 51 percent reduction from stormwater runoff.[6]

Before redevelopment, 80 percent of the 12-acre mill complex was covered with impervious surfaces, including buildings and pavement. Its stormwater drainage system was more than 100 years old, consisting of an ad hoc assemblage of catch basins and pipes that discharged untreated stormwater directly to the Charles River. Many of the pipes were routed under existing buildings and were broken or plugged.[7] In addition, the site offered no public access to the river.[8] Contamination due to its past industrial uses

Exhibit 3. The Waltham Watch Factory sits on the bank of the Charles River. Its redevelopment gives people access to the river.

created challenges for using green infrastructure, but the antiquated drainage system and the project's proximity to the Charles River made it important to reduce phosphorus loads and improve water quality.

B.   PLANNING AND REGULATORY CONTEXT

The project required a special permit under the city's Riverfront Overlay District.[9] The city established the district to guide the redevelopment of land along the Charles River. The overlay district is meant to promote development compatible with a riverfront setting and increase public holdings, public views of, and public access to the river. However, the overlay district does not address stormwater issues or the use of green infrastructure.

Although the city did not require green infrastructure at the project site, Watch City Ventures recognized that the river was a signature asset for its waterfront property, and water quality improvement in the river necessitated reducing stormwater flows to the river. Therefore, Watch City Ventures made a

[4] Charles River Watershed Association and Numeric Environmental Services, Inc. *Total Maximum Daily Load for Nutrients in the Upper/Middle Charles River, Massachusetts.* 2011. http://www.mass.gov/eea/docs/dep/water/resources/n-thru-y/ucharles.pdf.
[5] Charles River Watershed Association. "Charles River Watershed." http://www.crwa.org/charles-river-watershed. Accessed Apr. 29, 2015.
[6] Charles River Watershed Association and Numeric Environmental Services, Inc. *op. cit.*
[7] Reed, Peter, and Kate Bowditch. *The Watch Factory: A Case Study in Low Impact Development and Community*

Involvement. 21st Annual Nonpoint Source Pollution Conference. May 17-19, 2010. https://www.neiwpcc.org/npsconference/10-presentations/Reed%20and%20Bowditch%20-%20Watch%20Factory.pdf.
[8] Charles River Watershed Association. "Waltham Watch Factory." http://www.crwa.org/blue-cities/demonstration-projects/waltham-watch-factory. Accessed Apr. 29, 2015.
[9] City of Waltham. "Zoning Code. Sec. 8.4 Riverfront Overlay District special permit (RF)." Amended Dec. 9, 1991. http://ecode360.com/26938380.

commitment early in the design process to include green infrastructure practices and invited the Charles River Watershed Association to join the development team as an independent consultant to evaluate plans, propose alternatives, and provide feedback on the project.[10] City officials liked the green infrastructure plans that Watch City Ventures presented during the permitting process so much that the zoning board made implementing those plans a condition for approving the project.

C. DESIGN AND PERFORMANCE

The project's stormwater discharge into the Charles River required adherence to the Massachusetts Stormwater Management Standards[11] and a permit from the local conservation commission. The standards require redevelopment projects to meet or exceed 10 performance standards to the maximum extent practicable and, more importantly, to demonstrate improvement over existing conditions.

The site design, per recommendations from the Charles River Watershed Association, also focused on reducing the temperature and nutrient levels of stormwater runoff (Exhibit 4). Overall, the site was designed to:

- Treat the first inch of stormwater runoff from any impervious surfaces to reduce phosphorus levels and improve water quality.

- Recharge groundwater with up to 0.6 inches of runoff from impervious surfaces to the maximum extent practicable, conforming to the Massachusetts Department of

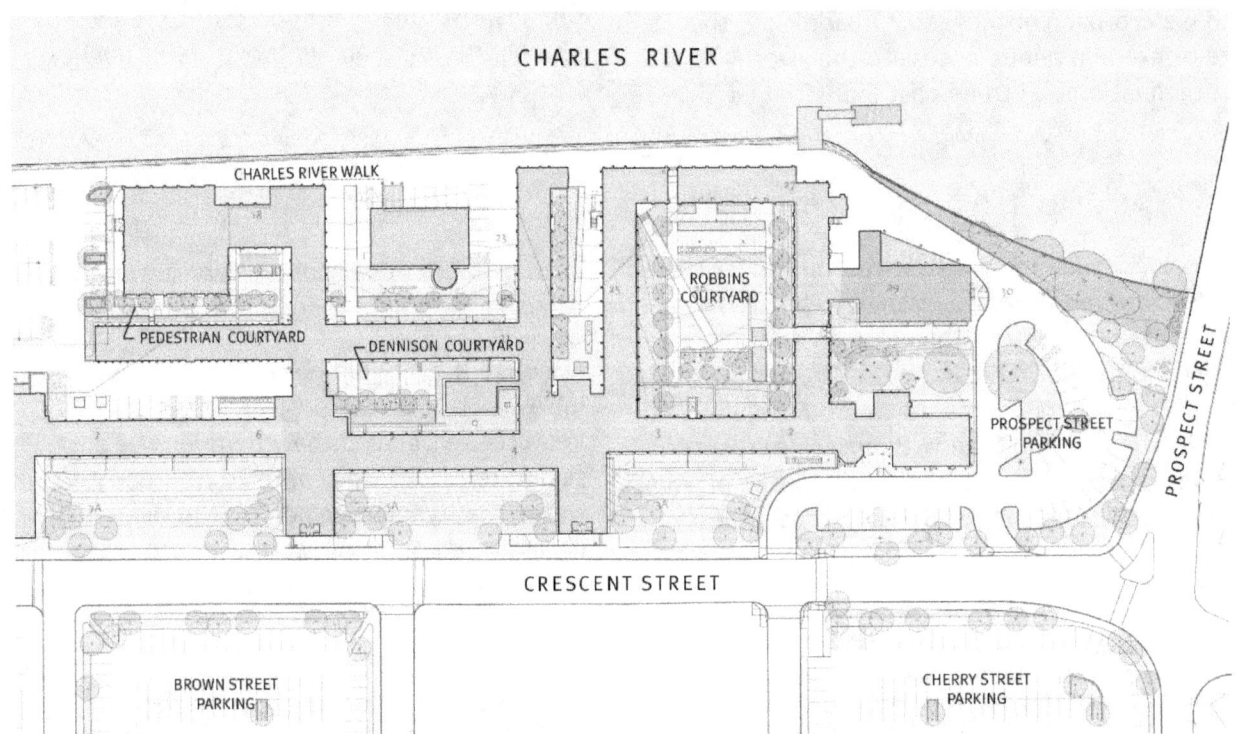

Exhibit 4. The Waltham Watch Factory site plan shows how landscaped spaces were integrated throughout the project to reduce overall impervious area.

---

[10] Charles River Watershed Association, "Waltham Watch Factory" op. cit.
[11] Commonwealth of Massachusetts. Massachusetts Stormwater Handbook. 2008. http://www.mass.gov/eea/ agencies/massdep/water/regulations/massachusetts-stormwater-handbook.html.

Exhibit 5. Dennison Courtyard before and after redevelopment shows how new canopy trees, rain gardens, and tables make the space functional and inviting.

Environmental Protection groundwater recharge criteria.

- Minimize the speed and volume of stormwater runoff from the discharge point along the river.

Phase 1 met these design objectives through a reduction of total site impervious area. At the exterior parking lots along Crescent Avenue, stormwater passes over a grass filter strip that drains into deep infiltration trenches to provide pretreatment and ground water recharge, reducing runoff to the municipal stormwater drainage system. Along the exterior perimeter of the existing mill buildings, the old pavement was replaced with grass filter strips and infiltration trenches that capture roof and surface runoff. Finally, rain gardens in the interior courtyards filter runoff from the surrounding roofs and courtyard paving (Exhibit 5). The rain gardens are lined with an impermeable geomembrane because soil contamination in this area precludes infiltration.

Overflow structures for all three locations convey runoff from larger storm events into a closed pipe system that discharges into the Charles River. The site is designed to reduce peak runoff volume between 5 and 9 percent,

depending on the size of the storm.[12] Phase 2 involved using porous asphalt to reduce flow, volume, water temperature, and nutrient loading at the Prospect Street parking lot.

High groundwater and soil contamination limited the use of green infrastructure in the parking area along the river. Instead, designers used more conventional drainage structures with oil-water separators and hydrodynamic separators that remove sediment and other pollutants.

After three years, visual site inspections during both dry and wet periods indicated all of the green infrastructure practices were meeting or exceeding expected infiltration rates. In addition, freezing and sanding of the parking lots during the winter do not appear to have affected performance.[13] Testing conducted during a storm in 2013 found that discharge from the two courtyard rain gardens had 30 to 50 percent less nitrate, 30 to 40 percent less phosphate, and 60 percent more dissolved oxygen than stormwater on-site that did not flow through the rain gardens.[14] The Prospect Street parking lot courtyard, with its preserved mature shade trees, was 13°F cooler in the summer than the Robbins courtyard, which does not have mature trees.[15]

[12] Landscape Architecture Foundation. "Watch Factory, Phases 1 and 2." http://landscapeperformance.org/case-study-briefs/watch-factory. Accessed Apr. 30, 2015.

[13] Personal communication with Kate Bowditch, Charles River Watershed Association, and Eric Ekman, Berkeley Investments, Inc., on Apr. 5, 2011.
[14] Landscape Architecture Foundation *op. cit.*
[15] Ibid.

D. COSTS AND FUNDING

The total construction budget for phase 1 of the project was $25 million.[16] Federal and state historic tax credits, state brownfield redevelopment tax credits, and a historic tax credit bridge loan supplemented private funding.[17] Stormwater management for phase 1 was $434,600, or 2 percent of total project costs (Exhibit 6). Green infrastructure is often incorporated into other site design features, making it difficult to isolate its capital costs from the overall construction budget. For example, the rain gardens were larger (and more expensive) than necessary to manage stormwater because designers wanted to make them attractive amenities.

STORMWATER MANAGEMENT PRACTICE	QUANTITY	UNIT COST	COST
Brown Street lot infiltration trench	220	$161 per linear foot	$35,400
Brown Street lot tree well with tree	2	$6,050 each	$12,100
Building 4 front courtyard infiltration trench	135	$142 per linear foot	$19,200
Pedestrian Courtyard rain garden	270	$622 per square foot	$167,900
Dennison Courtyard rain garden (including all landscape elements)	700	$286 per square foot	$200,000
Total			$434,600

Exhibit 6. Completed green infrastructure cost summary.

Source: Columbia Construction Company

E. BENEFITS

The developer of the Waltham Watch Factory recognized that the adjacent Charles River provides a valuable amenity that helps attract residents and businesses to the project. Using green infrastructure to improve the river's water quality thus helps to protect one of the site's inherent assets. The various green infrastructure features filter pollution and reduce stormwater flows into the river. A reduction of impervious surface area along the river also helps improve the stream bank and its value as a natural habitat. The project achieved additional environmental benefits by cleaning up a contaminated site and reducing ambient air temperatures during hot weather.

The green infrastructure at the Watch Factory not only helps improve the Charles River but also is an amenity for building tenants and visitors who enjoy the refurbished courtyards. The project helped to increase public awareness of

Exhibit 7. The Waltham Watch Factory redevelopment created an amenity for residents and visitors who can now enjoy views of and access to the Charles River.

the Charles River and the role green infrastructure can play to help protect it. Boardwalks and pedestrian paths now allow the public to reach the river's edge, increasing the value residents place on this vital asset for the city.

---

[16] Bruner/Cott. "The Watch Factory." http://brunercott.com /Project_shts_round_03/Project%20Sheets%20PDF_Commerci al/BC_proj_sht_watch_factory.pdf. Accessed Sep. 2, 2015.

[17] Watch City Ventures LLC. "Team." http://www.waltham watchfactory.com/team. Accessed Sep. 15, 2015.

F. LESSONS LEARNED

- Screening for soil contamination that could limit the ability to infiltrate stormwater should occur before project design to identify areas suitable for green infrastructure. Designers had to reconfigure the project after initial planning, reducing the expected benefits.

- Green infrastructure can be effectively integrated into an historic property, enhancing a project's aesthetic appeal. The gardens and green space incorporated into the project help highlight the connections between the river as a valuable community asset and development that protects and preserves it.

G. PROJECT TEAM

- **Owner:** Watch City Ventures, LLC, a joint venture between Berkeley Investments, Inc., and The First Republic Corporation of America
- **Developer:** Berkeley Investments, Inc.
- **General contractor (phase 1):** Columbia Construction Company
- **Architect:** Bruner/Cott & Associates
- **Landscape architect:** Richard Burck Associates, Inc.
- **Civil engineer:** BSC Group, Inc.
- **Environmental and geotechnical engineer:** Haley Aldrich
- **Watershed advisor:** Charles River Watershed Association
- **Environmental consultant:** Pine & Swallow Associates

- **Historic resource consultant:** Epsilon Associates[18,19,20]

Richard Burk Associates and Richard Mandelkorn Photography

Exhibit 8. Green infrastructure in the Watch Factory courtyards provides an attractive view for residents and a place to gather.

[18] Reed and Bowditch *op. cit.*
[19] Eckman, Eric. "The Redevelopment of the Historic Waltham Watch Factory." *The Weathervane.* 2009. http://docizz.com/preview/63472744.html.

[20] Schneider, Jay W. "The Watch Factory, Waltham, Mass." *Building Design + Construction.* Oct. 14, 2010. http://www.bdcnetwork.com/watch-factory-waltham-mass.

III. QUEENS BOTANICAL GARDEN

FLUSHING, NEW YORK

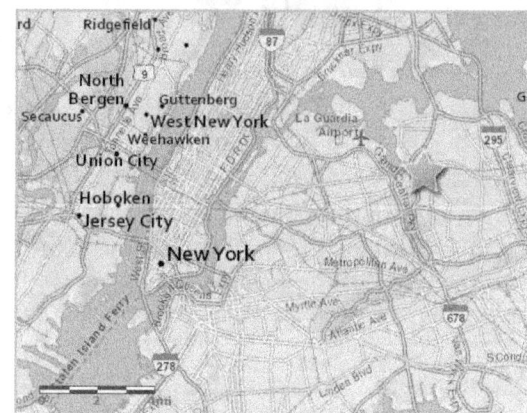

A new botanical garden visitor center and administration building demonstrates the feasibility of incorporating environmentally sustainable features into public buildings while educating visitors about stormwater management.

Project type:	Public building
Green infrastructure	Bioswales, cleansing biotope, green roof, artificial stream, and cisterns
Completion date:	2007

The Queens Botanical Garden's new visitor center and administration building is a $14 million public project that showcases sustainable design and green infrastructure. As a pilot project for the New York City Department of Design and Construction's High Performance Building Program, it paved the way for new green building requirements for city-funded projects. In 2008, it became the first building in New York City to earn a Leadership in Energy and Environmental Design (LEED) Platinum rating from the U.S. Green Building Council for new construction, providing a public example of successful sustainable design. One of the main project goals was to eliminate stormwater discharges to the city's combined sewer system. The project uses a combination of green infrastructure practices that both educate visitors and manage and treat stormwater on-site. The building also integrates several other environmentally sustainable practices, including grey water[21] recycling through a constructed wetland, stormwater infiltration through permeable pavers in the parking area, and renewable energy.

---

[21] Grey water is wastewater generated from building fixtures not including toilets—in this case, sinks, dishwashers, and showers.

A. SITE CONTEXT

The Queens Botanical Garden is in Flushing, Queens, in New York City (Exhibit 9). From the time of European settlement in the 1600s through the 1800s, the area was primarily agricultural, with the Flushing River providing a route for shipping products to market. However, in the 20th century, the area became a dense residential and commercial area with new bridge, subway, and rail connections to Manhattan.[22]

The site was a construction dumping ground for two World's Fairs. Soils at the site consist of urban fill covered by approximately 6 feet of imported soil. In addition, construction of the 1964 World's Fair filled in Mill Creek, a low-lying wetland that was a tributary to the Flushing River and once ran through the site.[23] The site was once a brownfield, but contamination from past uses has been remediated.

The surrounding neighborhood is a lower-income area with relatively few natural areas. The

Exhibit 9. The Queens Botanical Garden is in a dense residential and commercial area.

garden gives the neighborhood much-needed green space.

B. PLANNING AND REGULATORY PROCESS

The renovation of the Queens Botanical Garden visitor center and administration building was a pilot project for the New York City Department of Design and Construction's High Performance Buildings Program, which had the goal of making new and renovated public buildings in the city environmentally sustainable.[24] The process began in 1999 with a series of public workshops. They helped establish the design framework for the new building, which occupies approximately

4 acres in the 35-acre garden. Community members and staff helped identify water as a key element of the project that could help connect people with nature. Because clean, fresh water is important to cultures from around the world, the Botanical Garden wanted to illustrate the varied relationships people have to water for the diverse population that visits the garden, more than 75 percent of whom speak a language other than English at home.[25]

C. DESIGN AND PERFORMANCE

The garden's 2002 master plan set the goal of achieving zero stormwater runoff to avoid any discharge to the city's combined sewer system.

This goal was more stringent than city regulations called for but in keeping with the Botanical Garden's vision of highlighting the

---

[22] Conservation Design Forum and Atelier Dreiseitl. *Master Plan.* Queens Botanical Garden. 2002. http://www.queens botanical.org/103498/sustainable/master_plan.
[23] Ibid.

[24] Bernstein, Fred A. "A Garden Blooms in Queens." *Metropolis Magazine.* 2008. http://www.metropolismag. com/February-2008/A-Garden-Blooms-in-Queens.
[25] Conservation Design Forum and Atelier Dreiseitl *op. cit.*

importance of sustainable water resource management in cities.[26]

Multiple best management practices ensure that stormwater is retained on-site for all but the largest storms (Exhibit 10). Practices include permeable and semi-permeable surfaces to reduce runoff volume and water storage and treatment facilities for what remains. The visitor center and administration building has a 3,000-square-foot green roof over half of the structure. Excess runoff from the other half of the roof and adjacent walkways flows to a cleansing biotope, a highly permeable three-layer substrate of sand, gravel, and mineral additives, where soil and roots from native plants filter stormwater. Filtered water collects in a basin and fills a 24,000 gallon cistern, which feeds a fountain at the Botanical Garden's main gate. From the fountain, water flows through an artificial

meandering stream back to the basin, creating a closed-loop system. Storm events that exceed the capacity of the cistern and stream overflow to a large bioswale where runoff infiltrates or evaporates. Only the largest storms overflow into the city's combined sewer system. The stream runs dry during times of very low precipitation, as a natural system would.[27]

A smaller stormwater cistern for washing vehicles and building maintenance was constructed in the service area in the back of the building. In addition, stormwater runoff from parking areas either infiltrates through permeable pavers or flows to adjacent bioswales.[28] Together, the stormwater management practices annually treat and/or infiltrate approximately 628,000 gallons of runoff from 34,000 square feet of contributing area, 74 percent of which is impervious cover.[29]

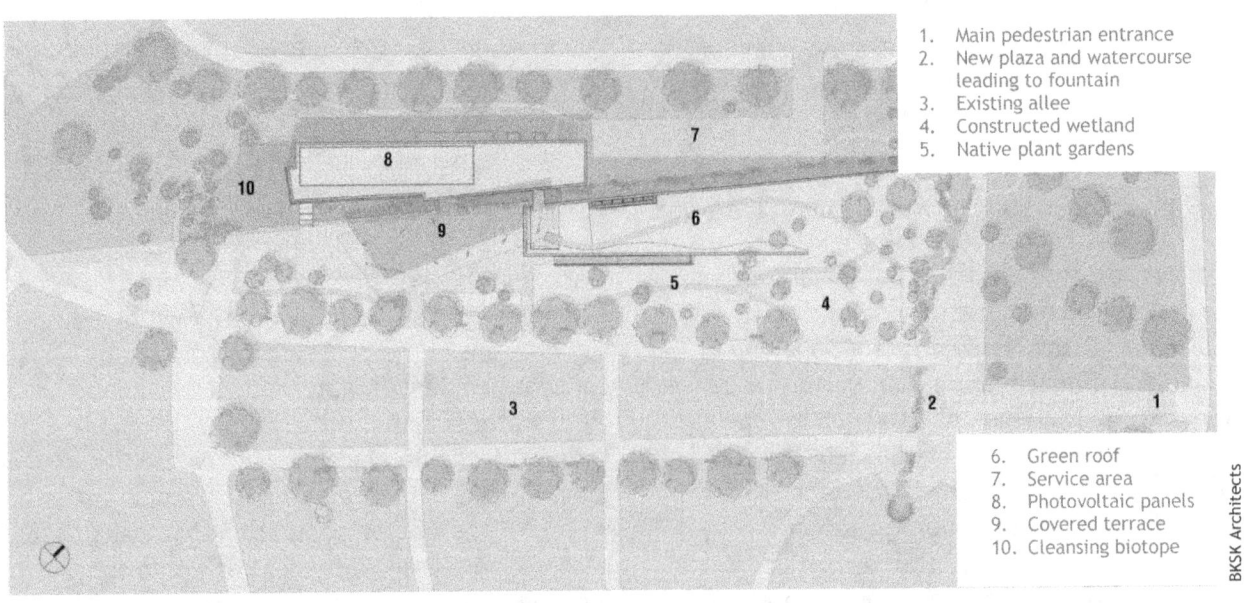

1. Main pedestrian entrance
2. New plaza and watercourse leading to fountain
3. Existing allee
4. Constructed wetland
5. Native plant gardens

6. Green roof
7. Service area
8. Photovoltaic panels
9. Covered terrace
10. Cleansing biotope

BKSK Architects

Exhibit 10. Queens Botanical Garden visitor center and administration building site plan.

D.   COSTS AND FUNDING

Project costs for stormwater management are difficult to isolate because water resource management and associated public education were integral components of the entire project.

The total cost for the visitor center and administration building was $14 million,[30] with

[26] Ibid.
[27] Bernstein *op. cit.*
[28] Conservation Design Forum and Atelier Dreiseitl *op. cit.*

[29] Atelier Dreiseitl. *Rainwater System Overview.* 2003.
[30] Bernstein *op. cit.*

the stormwater features accounting for about $568,000, or 4 percent of total costs.[31]

Most of the project was funded by the Office of the Borough President of Queens, with additional support from the Office of the Mayor, the New York City Council, several state agencies, and several foundations.

E. BENEFITS

By serving as a pilot project under New York City's High Performance Buildings Program, the Queens Botanical Garden helped establish the feasibility of environmentally sustainable public buildings and pave the way for future legislation requiring new construction funded by the city to achieve a minimum LEED certification level.[32] The project integrates environmental education into its design, helping to demonstrate to its hundreds of thousands of annual visitors the importance of responsible stewardship of water resources. The building and outdoor spaces are used for professional and school group tours, public programs, festivals, private event rentals, local club meetings, community board meetings, and as a voting location during elections.[33] In addition, the project helped the garden establish and run a grant-funded green jobs training program in partnership with LaGuardia Community College.[34] The program prepares graduates for careers in green cleaning and waste management and building operations and maintenance.[35]

Stormwater runoff to the combined sewer system from the site has been eliminated for all but the largest storms, preventing more than 600,000 gallons from entering the system annually.[36] Instead, stormwater irrigates the Botanical Garden's plants and helps create a natural environment for the public to enjoy in a lower-income neighborhood with relatively few natural areas.

Exhibit 11. The space outside the administration building hosts performances and festivals.

F. LESSONS LEARNED

- Pilot projects can test the effectiveness of stormwater management practices and introduce new concepts to the design community, regulators, and the public, ultimately leading to policy changes. The Queens Botanical Garden demonstrated the feasibility of incorporating environmental sustainability into public buildings, leading to new city requirements.

- Green infrastructure can be a focal point of a building's design, helping to reestablish long-severed connections between the natural environment and residents.

[31] Personal communication with Julie Nelson, BKSK Architects, on Apr. 20, 2011.
[32] City of New York. "Local Law 86 Basics." http://www.nyc.gov/html/oec/html/green/ll86_basics.shtml. Accessed Aug. 18, 2015.

[33] Personal communication with Julie Nelson, BKSK Architects, on Nov. 18, 2015.
[34] Ibid.
[35] Green Jobs Training Program. "Home." http://greenworkforcenyc.org. Accessed Nov. 20, 2015.
[36] Atelier Dreiseitl op. cit.

G. PROJECT TEAM

- **Owner:** Queens Botanical Garden
- **Developer:** City of New York Department of Design and Construction
- **Landscape and water design:** Atelier Dreiseitl
- **Landscape architect:** Conservation Design Forum
- **Architect:** BKSK Architects
- **Civil and structural engineers:** Weidlinger Associates
- **Mechanical, electrical, and plumbing engineers:** P.A. Collins, PE[37]

BKSK Architects

Exhibit 12. The covered terrace where people gather overlooks the cleansing biotope.

[37] Queens Botanical Garden. "Project Team." http://www.queensbotanical.org/103498/sustainable/ ParkingGarden_project/project_team. Accessed Aug. 17, 2015.

IV. KENSINGTON CREATIVE AND PERFORMING ARTS HIGH SCHOOL

PHILADELPHIA, PENNSYLVANIA

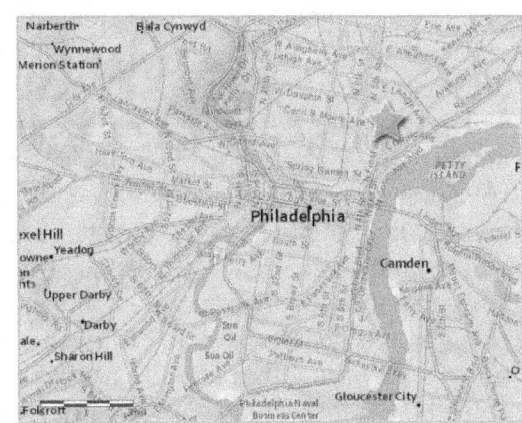

A new public high school on a former brownfield site becomes a showcase for environmental sustainability and a valued asset for the entire community.

Project type:	Public building; brownfield redevelopment
Green infrastructure practices:	Rain gardens, green roof, permeable pavement, rainwater cisterns, vegetative filter strips, and underground detention basins
Completion date:	Opened September 2010

Beginning in 2002, the nonprofit organization Youth United for Change began a push to break up the 1,400-student Kensington High School into four smaller schools that could be more responsive to student needs.[38] One of those schools would become the Kensington Creative and Performing Arts (KCAPA) High School in 2005. Students, parents, and community members successfully advocated for a new school building that would be a model of environmental sustainability.

In 2010, the KCAPA High School opened as the first public high school in the United States to be certified LEED Platinum.[39] Rain gardens, a green roof, porous pavement, vegetative filter strips, rainwater cisterns, and underground detention basins capture all stormwater on-site for reuse in irrigation and toilet flushing. Redevelopment of the 7.2-acre, formerly contaminated site turned a dangerous eyesore into a green amenity for the neighborhood and spurred redevelopment that incorporates green infrastructure on adjacent properties.

[38] Klonsky, Joanna. "Youth United for Change Takes on Philadelphia's Public Schools." What Kids Can Do, Inc. http://whatkidscando.org/featurestories/040107_YUC/index.html. Accessed Apr. 16, 2015.

[39] Delaware Valley Green Building Council. "Kensington High School for the Creative and Performing Arts." http://www.dvgbc.org/green_resources/projects/kensington-high-school-creative-and-performing-arts. Accessed Apr. 28, 2015.

A. SITE CONTEXT

The project site is between the revitalizing Fishtown neighborhood and the working-class, industrial neighborhood of South Kensington. The community wanted the school to help bring these neighborhoods together by creating a center where all could gather. The school district chose not to fence in the school,[40] allowing the community to get to the gym, cafeteria, and auditorium directly from the outside. Separate lobbies and mechanical systems facilitate after-hours use. Green infrastructure helped meet the community's goal to have an environmentally sustainable school that also is a neighborhood amenity. Extensive native plantings incorporated into stormwater management practices make the front of the school look like a neighborhood park and help make the school welcoming to residents.[41]

The school's 7.2-acre site is a narrow lot alongside a noisy elevated railway, and it presented several challenges for redevelopment. Contamination with lead, arsenic, and polyaromatic hydrocarbons from past industrial uses, including a former rail depot, had to be cleaned up.[42] The site had essentially been abandoned, attracting homeless people, drug dealers, and stray dogs.[43] It had the reputation of a dangerous place, and the community wanted to improve it.

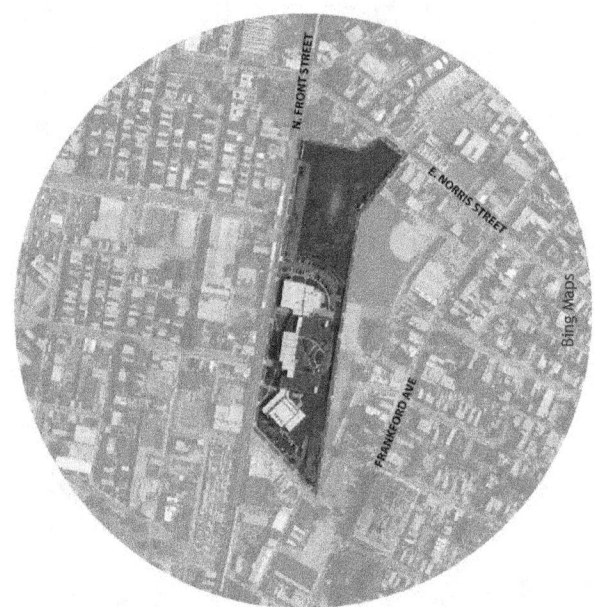

Exhibit 13. The KCAPA project site is a narrow lot surrounded by compactly developed neighborhoods.

B. PLANNING AND REGULATORY CONTEXT

In 2009, the Philadelphia Water Department launched Green City, Clean Waters, a plan to invest $2 billion in green infrastructure over 25 years to manage stormwater and protect the area's watersheds while revitalizing the city and achieving other environmental, social, and economic benefits.[44] Schools make up 2 percent of all impervious cover in the city's combined sewer service areas, and their high visibility makes them good opportunities for educating the community about the benefits of green infrastructure.[45] The Philadelphia Water Department helped advance the KCAPA project as one of the first in the city's Green Schools program, a component of Green City, Clean Waters.

Redevelopment projects disturbing more than 15,000 square feet of land must comply with the city of Philadelphia's 2006 stormwater management regulations that set requirements for water quality, channel protection, flood

[40] Ibid.
[41] ArchDaily. "The Kensington Creative and Performing Arts High School / SMP Architects and SRK Architects." Nov. 30, 2011. http://www.archdaily.com/187671/the-kensington-creative-and-performing-arts-high-school-smp-architects-and-srk-architects.
[42] Rath, Jane, and Travis Alderson. "Champion for Change." *High Performing Buildings.* Winter 2013. pp. 6-18. http://www.hpbmagazine.org/Case-Studies/Kensington-High-School-for-the-Creative-and-Performing-Arts-Philadelphia-PA.

[43] American Institute of Architects. "Kensington High School for the Creative and Performing Arts." http://www.aiatopten.org/node/48. Accessed Apr. 29, 2015.
[44] City of Philadelphia. "Target 8: Manage Stormwater To Meet Federal Standards." http://www.phila.gov/green/2011-progress-report/equity-target8.html. Accessed Apr. 20, 2015.
[45] Philadelphia Water Department. *Amended Clean City, Clean Waters: The City of Philadelphia's Program for Combined Sewer Overflow Control.* 2011. http://www.phillywatersheds.org/doc/GCCW_AmendedJune2011_LOWRES-web.pdf.

control, and non-structural site design.[46] However, this project was exempt from the channel protection requirement because it is in the Delaware River watershed.[47] Exhibit 14 summarizes the relevant requirements and the stormwater practices used to help meet each requirement.

REQUIREMENT	DESCRIPTION	PRACTICES IMPLEMENTED
Water quality	To recharge groundwater, restore natural site hydrology, reduce pollution in runoff, and reduce combined sewer overflows, the first inch of rainfall must be infiltrated on-site. Where infiltration is not appropriate or feasible, 100 percent of the runoff from directly connected impervious surfaces must be slowly released into the sewer system with 20 percent routed through an approved stormwater management practice to improve water quality.	• Porous pavement • Rain gardens • Green roof • Vegetative filter strips • Water quality inflow structures leading to underground storage
Flood control	To reduce flooding downstream of the development site and to reduce combined sewer overflows, peak runoff after development must not exceed peak runoff before development. Exact requirements depend on the flood management district in which the project is located. This project was required to reduce the 2-year storm event post-development peak flow rate to less than the 1-year storm event pre-development peak flow rate, and maintain post-development rates below pre-development rates for the remaining storms.	• Porous pavement • Rain gardens • Green roof • Underground detention
Public health and safety	To limit discharges to the combined sewer system, which has limited capacity, each of the sub-drainage areas must have a release rate no greater than 0.35 cubic feet per second per acre for up to the 10-year storm.	• Porous pavement • Rain gardens • Green roof • Vegetative filter strips • Underground detention
Non-structural site design	To reduce the quantity of stormwater that must be managed, projects must minimize creation of impervious cover and protect and use existing site features with natural stormwater management value.	• Efficient walkway layout • Smaller building footprint • Grass-turf pavers in loading and emergency access areas

Exhibit 14. Philadelphia stormwater regulations applicable to the project.

C. DESIGN AND PERFORMANCE

Soil contamination on the site made incorporating sufficient green infrastructure to meet stormwater management requirements challenging. Designers had to avoid infiltration in certain areas, but they maximized infiltration where appropriate. Field infiltration tests helped identify areas with soil conditions most favorable for green infrastructure practices. Measured infiltration rates ranged from 0.13 to 8.25 inches per hour at varying depths. The lowest rate used for infiltration was 2.25 inches per hour at one of the porous pavement areas. The rain gardens were sited in an area with an infiltration rate of 4.65 inches per hour. Soils near the playing fields and the fire lane had poor infiltration rates, so underground storage was used with water quality inflow structures to pretreat the runoff. Exhibit 15 provides the amount of directly connected impervious area (DCIA) drained, the amount of water treated (WQv), and the storage volume for each type of practice.

[46] City of Philadelphia. *Stormwater Management Program.* 2007. http://www.phillywatersheds.org/doc/2007_Annual_Report_Final.pdf.

[47] City of Philadelphia. *Stormwater Management Guidance Manual Version 2.1.* 2014. http://www.pwdplanreview.org/manual-info/pre-july-2015-resources.

STORMWATER PRACTICE	TYPE OF PRACTICE	TOTAL DCIA (SQUARE FEET)	WQV (CUBIC FEET)		STORAGE (CUBIC FEET)
			REQUIRED	PROVIDED	
Porous pavement	Infiltration	40,730	3,393	5,482	24,706
Pretreatment and underground storage	Treat and release	66,080	5,507	5,507	41,660
Rain gardens	Infiltration	8,580	715	2,193	4,507
Green roof	Retention/evapotranspiration	22,040	824	824	4,010
Total		137,430	10,439	14,006	74,883

Exhibit 15. Stormwater best management practices performance.

All infiltration practices were designed to handle their entire contributing drainage area for the 100-year storm, exceeding the requirements for public health and safety in combined sewer areas. Rain gardens were designed to intercept runoff from landscaped areas as well. Finally, the underground storage reduces post-development peak flow rates by 74 percent, 54 percent, and 33 percent for the 15-, 50-, and 100-year storms, respectively, compared to pre-development peak flows.

Two underground rainwater cisterns provide water for toilet flushing, although the seasonality of high school occupancy limits their effectiveness in managing stormwater runoff. Overflow from these cisterns is directed to one of the underground detention basins.

Because the project engineers worked closely with the Philadelphia Water Department on the design for this project, they were able to use new piping materials that the old code did not initially allow, resulting in cost savings and a better design.[48]

The school opened to students in September 2010. Five years after installation, the green infrastructure practices continue to function well. The permeable pavement surfaces are in good condition, and the vegetation is healthy.[49] Routine school maintenance incorporates upkeep of the green infrastructure practices along with more conventional landscaped areas.[50]

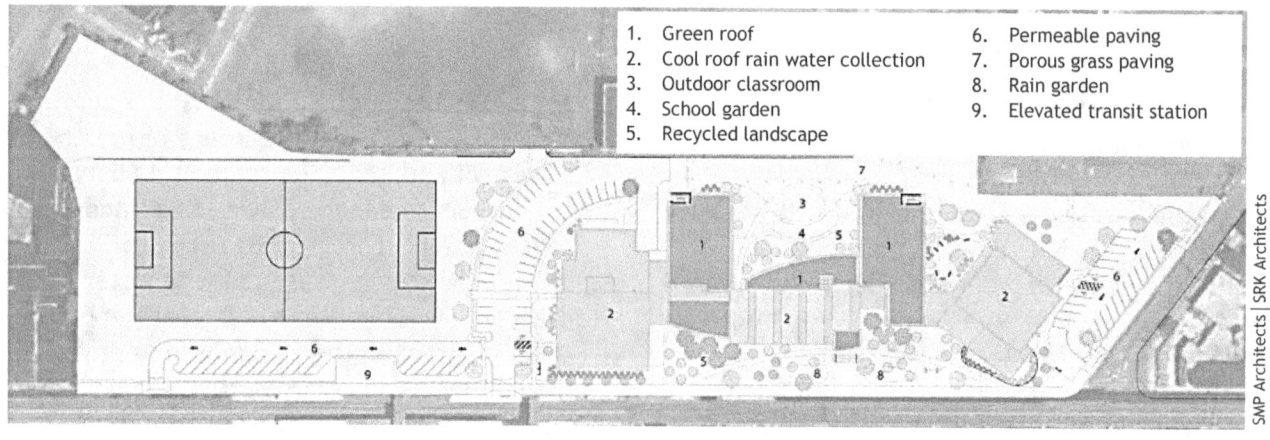

1. Green roof
2. Cool roof rain water collection
3. Outdoor classroom
4. School garden
5. Recycled landscape
6. Permeable paving
7. Porous grass paving
8. Rain garden
9. Elevated transit station

SMP Architects | SRK Architects

Exhibit 16. The site plan for KCAPA incorporated green infrastructure in multiple places.

[48] American Institute of Architects *op. cit.*
[49] Personal communication with Ronald Monkres, Gilmore & Associates, on Oct. 20, 2015.
[50] Personal communication with Ronald Monkres and Chris Green, Gilmore & Associates, on Feb. 11, 2011.

D. COSTS AND FUNDING

The School District of Philadelphia chose to construct KCAPA as a "turnkey" project, wherein a private developer purchased and cleaned up the land, built the school, and financed most of it through a local bank, turning the project over to the school district at completion for a set price.[51] The funding source for the $44 million project[52] was initially private, but the project later became a public-private partnership with the infusion of public funds, including $1 million from the Pennsylvania Department of Environmental Protection for a geothermal heating and cooling system.[53] The stormwater management practices cost about $1 million, or about 2.5 percent of total project costs.[54]

E. BENEFITS

Managing stormwater on-site with a combination of green infrastructure and underground storage reduces runoff and pollution entering the area's combined sewer system. The reduction in impervious cover reduces the site's stormwater fees, and rainwater harvesting and low-flow plumbing fixtures reduced projected municipal water use by 65 percent, with no water required for irrigating landscaping.[55] Overall, 69 percent of the site is green space due to compact building design and a geothermal HVAC system, which reduced the space needed for the mechanical systems.[56]

Exhibit 17. Riders of the elevated rail line can see the rain gardens in front of KCAPA as they pass by daily.

The environmental benefits of the school's green infrastructure extend beyond the site because it generated interest in expanding the practices used at the school to other places in the community. The New Kensington Community Development Corporation led a project to add green infrastructure to an adjacent recreation center, and that in turn led to green infrastructure improvements along the neighboring street.[57] In addition, the school's green roof and rain gardens are visible to the thousands of commuters passing by daily on the elevated rail line, helping to demonstrate the aesthetic value of green infrastructure to the community.

The school's location also created environmental benefits independent of the building design. Not only is a former contaminated industrial site now cleaned up, creating space for children's sports and community gardens,[58] but transportation-related emissions are low. More than 95 percent of building occupants travel to school by walking, biking, or taking public transportation, including the elevated rail line that runs along the school property. The community uses the elevated rail

[51] American Institute of Architects *op. cit.*
[52] BSI Construction. "Kensington High School for the Creative and Performing Arts." http://www.bsiconst.com/projects/case-studies/kensington-capa. Accessed Sep. 2, 2015.
[53] Rubin, Daniel. "Philadelphia District Wins a Green-Schools Award." *Philly.com*. Dec. 12, 2011. http://articles.philly.com/2011-12-12/news/30507242_1_green-schools-green-roofs-green-cleaning-products.

[54] Personal communication with Ronald Monkres, Gilmore & Associates, on Oct. 19, 2015.
[55] Rath and Alderson *op. cit.*
[56] Eco-structure Staff. "Kensington High School for the Creative and Performing Arts." *Ecobuilding Pulse*. Aug. 14, 2012. http://www.ecobuildingpulse.com/award-winners/cote-2012-top-ten-kensington-high.aspx.
[57] Rath and Alderson *op. cit.*
[58] American Institute of Architects *op. cit.*

stop more, helping to revitalize the neighborhood.[59]

Student engagement and performance at KCAPA improved after moving to the new building. In the first year of operation, the school had a waiting list to attend, while truancy rates fell and both test scores and the graduation rate went up.[60] School staff have also noticed less littering and vandalism at the new school as students take pride in the building and become stewards of the space.[61]

F. LESSONS LEARNED

- Green infrastructure can be used successfully on a contaminated site where only part of the site is suitable for infiltration. Careful site design based on soil testing allowed designers to maximize infiltration by placing green infrastructure in the most suitable locations.

- Less than one-third of the parking built at the school (required by zoning) is being used. The developers did not pursue a zoning variance because of the time required. The school could have used the half-acre of unneeded parking space for additional green space or other uses if Philadelphia's zoning required less parking for properties near public transit in walkable neighborhoods.[62]

- Often, school districts feel pressure to eliminate environmentally sustainable features when budgets are tight, but constructing the project with a "turnkey" delivery allowed the developer to use sustainable features as long as it could meet the overall budget and schedule. Constructing the school with a smaller footprint ultimately saved enough money to pay for the green features.[63]

G. PROJECT TEAM

- **Owner:** The School District of Philadelphia
- **Design:** SMP Architects and SRK Architects
- **Engineering:** Gilmore & Associates
- **Construction:** AP Construction and Bustleton Services, Inc.[64]

[59] Rath and Alderson *op. cit.*
[60] Eco-structure Staff *op. cit.*
[61] Personal communication with Ronald Monkres, Gilmore & Associates, on Oct. 19, 2015.
[62] Rath and Alderson *op. cit.*
[63] American Institute of Architects *op. cit.*
[64] Stabert, Lee. "Learning Curve." *Grid*. Nov. 2010. http://issuu.com/redflagmedia/docs/grid_2010.11?e=126199 5/5375442#search.

V. THE RADIAN
COMPLEX

PHILADELPHIA, PENNSYLVANIA

A new mixed-use building on a university campus uses a green roof system to meet stormwater requirements while maximizing developable space, creating an amenity for residents and businesses.

Project type:	Mixed-use development
Green infrastructure practices:	Green roof, pervious pavers, tree pits and planters, and stormwater detention basins
Completion date:	2009

The University of Pennsylvania and a private developer built a new mixed-use building with retail and student apartments. The site at the edge of the University's Philadelphia campus had existing structures and was 99 percent impervious. Using green infrastructure, they achieved a 30 percent reduction in impervious area draining to the combined sewer system.

New trees and a green roof are now selling points for the commercial and residential tenants. Although the site is highly constrained, the developer was able to meet stormwater management requirements for green infrastructure while maximizing the developable area of the site.

A. SITE CONTEXT

The Radian Complex is a 14-story, 500-bed student housing and retail center on Walnut Street at the northwestern edge of the University of Pennsylvania campus (Exhibit 18). The project is part of continuing campus development along the 40th Street corridor, which links the university to West Philadelphia. A project combining student housing and neighborhood-serving retail helps to link these two areas of the city.[65]

Before construction, the project site was 99 percent impervious, sending nearly all stormwater runoff into the city's combined sewer system. The site is in the Schuylkill River

[65] Wisniewski, Katherine. "Radian Apartments/Erdy McHenry Architecture." *Arch Daily.* Aug. 15, 2011.

http://www.archdaily.com/158386/radian-apartments-erdy-mchenry-architecture.

watershed. Reducing combined sewer overflows into the Schuylkill River is a critical step for meeting water quality standards and keeping aquatic life healthy.[66] However, other buildings and infrastructure near the project site limited the opportunity for infiltration.

B. PLANNING AND REGULATORY CONTEXT

On January 1, 2006, the city of Philadelphia Water Department instituted new stormwater management regulations, providing guidelines for achieving water quality and managing runoff.[67] This project was one of the first to have to meet the new requirements.

The regulations require redevelopment projects to reduce predevelopment runoff volumes by at least 20 percent based on the following:

- All non-forested, pervious areas are considered meadow (in "good" hydrologic condition).
- In addition, 20 percent of existing impervious cover on-site is also considered meadow.

Exhibit 18. The Radian Complex sits on the edge of the University of Pennsylvania's campus along a commercial corridor.

The site must also infiltrate on-site, or if soils do not permit, store and treat on-site the stormwater volume equal to the first inch of rainfall over all directly connected impervious areas.

C. DESIGN AND PERFORMANCE

To help meet the stormwater requirements while maximizing the ground-floor space available for retail, designers used a 12,000-square-foot green roof system that covers 17 percent of the site. In addition, 4,700 square feet of adjacent conventional roof area, covering an additional 6 percent of the site, drains to the green roof system, allowing the project to exceed the required 20 percent reduction in impervious cover.[68] The system includes five green roofs. Three extensive roofs are designed to hold the first inch of runoff. Two intensive roofs are designed to hold the first 2 inches of runoff.

Together, they treat the first inch of runoff for 16,700 square feet, meeting water quality requirements. The green roof system maximizes water retention on the roof and controls the release rate into two underground stormwater management basins to meet the city's channel protection and flood control requirements.

Even with the reduction in impervious cover afforded by the green roof, the project's design still had to manage stormwater runoff from the rest of the project site. The project's retail plaza incorporates interlocking permeable

---

[66] Philadelphia Water Department. *Green City Clean Waters.* 2011. http://www.phillywatersheds.org/doc/GCCW_AmendedJune2011_LOWRES-web.pdf.
[67] Philadelphia Water Department. *Stormwater Regulations,* §600.0 Stormwater Management. 2006.

[68] Pockl, Andrew, and Corey Fenwick. "More than Just a Pretty Garden." *Stormwater Solutions.* Nov./Dec. 2007. http://estormwater.com/More-Than-Just-a-Pretty-Garden-article8747.

concrete pavers and tree pits. Stormwater is directed first into planters that provide some water storage in a layer of stone below the soil. Excess water collects in an underdrain and flows through tree pits to one of two underground detention basins. The basins are constructed from "milk crate" type structures covered with gap-graded stone placed within an impermeable geotextile liner. The detention basins are designed to collect the overflow from a 100-year storm and slowly release it to the combined sewer system at a controlled rate.[69]

The project reduced directly connected impervious area by 30 percent using a combination of a green roof, pervious pavers, planters, and tree pits, as described in Exhibit 20. In addition, an outdoor dining space on the

Exhibit 19. The green roof system on the Radian Complex diverts stormwater from the combined sewer system while giving residents an attractive view.

upper terrace level with views of the street includes a grove of trees separating the retail and residential components.

STORMWATER MANAGEMENT PRACTICE	SQUARE FEET	PERCENT OF SITE
Green roof area	12,091 (650 intensive; 11,441 extensive)	17.8%
Conventional roof area draining to green roof	4,022	5.9%
Pervious pavers	2,155	3.2%
Planter boxes	1,470	2.2%
New trees	900	1.3%
Total	20,638	30.3%

Exhibit 20. Reduction in directly connected impervious area for stormwater management practices.

D. COSTS AND FUNDING

This project was privately funded on land owned by the University of Pennsylvania. Excluding the cost of landscaping, the green roof, porous pavement, planters, tree pits, and detention basins together cost $377,000, or 0.5 percent of the total project cost of $70.2 million (Exhibit 21). The green roof was the most costly element of the stormwater management system, but it allowed the project to meet stormwater management regulations while maximizing the area available for retail.

Annual operations and maintenance costs are estimated to be about $6,000 based on price quotes submitted to the engineers for the project. These costs include routine inspections of sumps in all inlets and roof drains for debris removal, inspection of outlet structures on the underground detention basins after all major storms for removal of silt and debris, inspection of pervious pavers, and sweeping to remove debris.

---

[69] Ibid.

FEATURE	QUANTITY	UNIT COST	COST
Green roof, including protection layer, drainage media, perforated pipes, soil, and plantings	12,000 ft²	$14 per ft²	$165,000
Pervious pavers, including site preparation, gravel base, and underdrain pipes	2,150 ft²	$25 per ft²	$53,500
Stormwater drainage system, including detention basins and piping	n/a	n/a	$158,500
Total			$377,000

Exhibit 21. Construction cost summary. Costs for the green roof do not include elements that are part of a conventional roof such as the roof structure and waterproofing, which add approximately $15 per square foot.
Source: Pennoni Associates.

E. BENEFITS

This project reduced the amount of impervious surface area at the project site by 30 percent, exceeding regulatory requirements and reducing runoff flowing to the city's combined sewer system. The primary component of the stormwater management system, the green roof, creates an attractive view for the student apartments and retail businesses that can see it, allowing the project owner to charge higher rent for the retail area overlooking the green roof. The tree planters on the sidewalk and courtyard next to the retail area create an additional aesthetic amenity for the neighborhood.

The use of a green roof allowed the owner to maximize the available space for retail while meeting stormwater management requirements, supporting the overall business environment in the neighborhood. The ground-level retail extends along the entire block on the 40th Street

Peter Kubilis/Erdy McHenry Architecture

Exhibit 22. Retail space on the first floor of the Radian Complex helps bring activity to the commercial corridor.

retail corridor, serving both the university and the adjacent residential neighborhood. Since the building opened, additional development has occurred in the immediate area, and restaurants and businesses are thriving.[70]

F. LESSONS LEARNED

- Even relatively expensive green infrastructure practices like green roofs can be economically viable when they allow project developers to meet stormwater requirements while maximizing developable area on a site. Higher costs with some green infrastructure practices can be offset by reduced construction and maintenance of conventional stormwater infrastructure and by the ability to command higher prices for the property due to the green infrastructure.

- Water quality improvements are possible even on properties in highly developed areas where soils are not conducive to infiltration. The Radian Complex used a green roof to meet requirements for managing stormwater on-site at a location where other buildings and infrastructure limited options.

[70] Personal communication with Andrew Pockl, Pennoni Associates, on Nov. 13, 2015.

G. PROJECT TEAM

- **Owner:** University of Pennsylvania.
- **Developer and manager:** University Partners
- **Civil engineers and landscape architects:** Pennoni Associates
- **Architects:** Erdy McHenry Architecture, LLC
- **Structural engineer:** The Harman Group
- **Green roof consultant:** Roofscapes
- **Exterior wall consultant:** Edwards & Company
- **Contractor:** INTECH Construction, Inc.[71]

---

[71] Erdy McHenry Architecture, LLC. "The Radian." *Architype Review*. 2008. http://architypereview.com/project/the-radian/?issue_id=561.

VI. SAND RIVER HEADWATERS GREEN INFRASTRUCTURE PROJECT

AIKEN, SOUTH CAROLINA

A downtown project incorporating green infrastructure into public spaces enhances a small town's historic charm while helping to preserve a beloved urban forest and restore a degraded river.

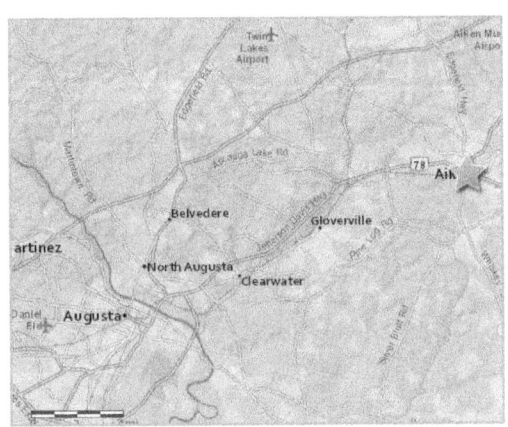

Project type:	Transportation; historic preservation
Green infrastructure practices:	Rain gardens, bioswales, porous asphalt, pervious concrete, permeable interlocking concrete pavers, and underground cisterns
Completion date:	2011

Over decades, stormwater eroded a 70-foot-deep canyon below the Aiken, South Carolina, stormwater outfall (Exhibit 23). The banks of the Sand River destabilized, destroying vegetation and choking downstream wetlands with sediment as they collapsed. The city chose a restoration plan for the Sand River that focuses on upstream reduction of stormwater runoff through green infrastructure in downtown Aiken, including bioswales, porous asphalt, permeable pavers, and rain gardens. The city chose this approach as the most cost-effective way to remedy environmental degradation that could also improve the city's historic parkways and boulevards with wide, landscaped medians.

Clemson University Intelligent River®

Exhibit 23. Stormwater has eroded a deep canyon in the banks of the Sand River.

A. SITE CONTEXT

Aiken is in western South Carolina near the Georgia border. The city of approximately 30,000 people has one of the largest urban forests in the country, a 2,100-acre public green space next to downtown called Hitchcock Woods (Exhibit 24). Through the forest flows the Sand River, an ephemeral stream that runs dry between periods of rain.

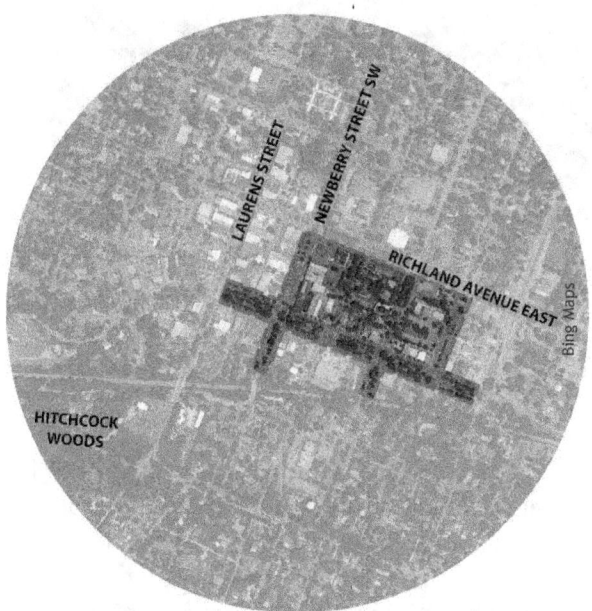

Exhibit 24. Aiken's downtown is next to Hitchcock Woods, which receives stormwater runoff from city streets.

Downtown Aiken has 105 acres of 150-foot-wide parkways and boulevards with landscaped medians, giving the city a distinctive charm and character. Stormwater runoff from the streets originally flowed through open channels and a relatively natural drainage system. That changed in the 1950s, when the city paved the streets and installed a conventional storm sewer system that conveys stormwater to an outfall at the headwaters of the Sand River.[72] As impervious cover has increased with new development, stormwater runoff has eroded a 70-foot-deep canyon in the Sand River, sending loose sand downstream where it degrades forested wetlands.[73] The Sand River has been listed as an impaired waterbody since 1998 for exceeding standards for fecal coliform bacteria, with stormwater runoff listed as a potential major source of contamination.[74]

B. PLANNING AND REGULATORY CONTEXT

In 2008, the city awarded the Clemson University Center for Watershed Excellence (Clemson) a grant to develop a river restoration master plan. Clemson convened a series of workshops and meetings with the city, community members, and other stakeholders to explore alternatives. The preferred approach for river restoration would cost approximately $16 million to $18 million, including remediation of the canyon and wetlands, new pipes to convey flow below the restored canyon, energy dissipation and storage devices, and tributary stabilization. The parties ultimately chose a strategy focused initially on reducing runoff in the watershed through green infrastructure to fix the root cause of the problem. The city hopes this approach will improve the flow conditions at the outfall and potentially reduce the overall cost of the downstream river restoration project.[75]

[72] Eidson, G.W., et al. "Sand River Headwaters Green Infrastructure Project, City of Aiken, South Carolina: A Collaborative Team Approach to Implementing Green Infrastructure Practices." Proceedings of the 2010 South Carolina Water Resources Conference. Oct. 13-14, 2010.
[73] Clemson University. *Sand River Headwaters Green Infrastructure Project*. 2013. http://media.clemson.edu /public/restoration/sand%20river/agi_finalreport_022113-web.pdf.

[74] South Carolina Department of Health and Environmental Control. *Total Maximum Daily Load Horse Creek (Hydrologic Unit Code: 03060106060, -030, -040 & -050) Stations SV-069, SV-072, SV-073 & SV-250 Fecal Coliform Bacteria*. 2005. https://www.scdhec.gov/HomeAndEnvironment/Docs/tmdl_horse.pdf.
[75] Clemson University 2013 *op. cit.*

C. DESIGN AND PERFORMANCE

The Sand River Headwaters Green Infrastructure Project includes multiple green infrastructure practices installed near the intersection of Park Avenue and Newberry Street in downtown Aiken.

City staff and residents were concerned that this project would affect the general appearance of the historic parkways and disturb existing mature trees. To address this concern, designers chose bioswales and rain gardens that complement the existing parkway landscaping, using native plants to blend in with the surroundings (Exhibit 25). Water from adjacent roads and sidewalks flows to the bioswales and rain gardens, which filter out nutrients and bacteria, reduce peak discharge flows, and recharge ground water. Street improvements include porous asphalt; pervious concrete; and permeable, interlocking pavers in the parking lanes, which infiltrate runoff from the adjacent road as well. Overflow systems direct excess runoff to the existing storm sewer system for large storm events. Underground cisterns at several locations store runoff for irrigation.[76]

In 2008, the city did not have regulations specifying minimum design requirements for green infrastructure practices. Therefore, the practices are sized to meet the minimum standard design criteria issued by the South Carolina Department of Health and Environmental Control in 2002, which include:

- Infiltration practices must capture and treat the first inch of runoff from the contributing impervious surfaces.
- Post-development peak flows from best management practices must not exceed pre-development discharge rates for the 2- and 10-year, 24-hour storm events.[77]

Clemson University Intelligent River®

Exhibit 25. Rain gardens in street medians help manage stormwater while providing attractive landscaping.

In addition to meeting these requirements, the green infrastructure practices infiltrate at least the 2-year, 24-hour storm event (approximately 3.7 inches) within 72 hours.[78]

The first phase of construction was completed in February 2011. Monitoring equipment measures the performance of the green infrastructure practices as part of the online Clemson University Intelligent River Program, which displays real-time data. The program is an educational tool that the community, educators, and designers can view online.[79] The data include baseline measurements of stormwater hydraulics before installation of the green infrastructure practices to evaluate their impact.

Preliminary monitoring results indicate that in many cases, the bioinfiltration practices are infiltrating all stormwater runoff and discharging none into the sewer system through the overflow pipes. However, the Sand River watershed overall showed no statistically significant improvement, likely because the surface area of the bioretention practices represented just 0.4 percent of the total watershed. All three types of permeable pavement worked as expected, with average infiltration rates adequate for stormwater management.[80]

[76] Ibid.
[77] South Carolina Department of Health and Environmental Control. *Standards for Stormwater Management and Sediment Reduction Regulation 72-300 thru 72-316.* 2002. https://www.scdhec.gov/Agency/docs/water-regs/r72-300.pdf.

[78] Clemson University 2013 *op. cit.*
[79] Clemson University. "Intelligent River Data Browser for Sand River, Aiken." https://www.intelligentriver.org/data?p=7. Accessed May 14, 2015.
[80] Clemson University 2013 *op. cit.*

D. COSTS AND FUNDING

The city of Aiken was awarded $3.34 million under the American Recovery and Reinvestment Act of 2009 through the Clean Water State Revolving Fund for design, construction, and post-construction monitoring of green infrastructure practices to control stormwater runoff into the Sand River, the first phase of the Sand River restoration master plan.[81] The city awarded two related grants to Clemson:

$293,187 for the design of the green infrastructure practices and $126,359 for a research and monitoring program.[82]

The city is responsible for the operation and maintenance of the green infrastructure practices installed as part of this project. The city pays for those costs through a stormwater utility fee assessed to city property owners.[83]

E. BENEFITS

The green infrastructure practices installed as part of the master plan for restoring Sand River complement the landscaping of the city's historic parkways and boulevards. The city plans to use similar approaches to improve stormwater management in other parkways in and around the city to revitalize neighborhoods and further lessen the amount of stormwater the city discharges to the Sand River headwaters.

The monitoring program established as part of this project helps educate residents and leaders in the city of Aiken and other communities, educators, designers, and the scientific community, providing valuable information on the design effectiveness of green infrastructure practices. In the first three years, more than 5,000 people viewed the website.[84]

F. LESSONS LEARNED

- Green infrastructure can complement and enhance historic city landscapes. Residents in Aiken were initially concerned about possible damage to mature trees that lined the city's parkways and boulevards, and they did not want to detract from the city's overall historic charm. Permeable pavement and careful design of bioswales allowed green infrastructure to enhance the city's aesthetics.

- Green infrastructure in historic downtowns can help protect natural areas that residents cherish. The Aiken green infrastructure project benefits Hitchcock Woods, a natural area next to downtown that serves as a city park and is central to the area's identity. After residents understood the link between downtown stormwater and the health of their local ecosystem, they supported green infrastructure as an innovative approach for environmental protection.

G. PROJECT TEAM

- **Owner:** City of Aiken
- **Engineering:** Clemson University Center for Watershed Excellence and Woolpert, Inc.

---

[81] Ibid.
[82] Greenville.com Community News. "Aiken, Clemson, EPA Kick Off Project to Make Stormwater 'Green'." http://www.greenville.com/news/epa0310.html. Accessed Sep. 3, 2015.

[83] City of Aiken. *Storm Water*. 2015. https://www.cityofaikensc.gov/wp-content/uploads/downloads/2015/01/brochure_stormwater.pdf.
[84] Clemson University 2013 *op. cit.*

VII. MENOMONEE VALLEY INDUSTRIAL CENTER

MILWAUKEE, WISCONSIN

A redeveloped industrial center helps restart a region's economic engine while creating a stormwater park that connects residents to the long-isolated Menomonee River.

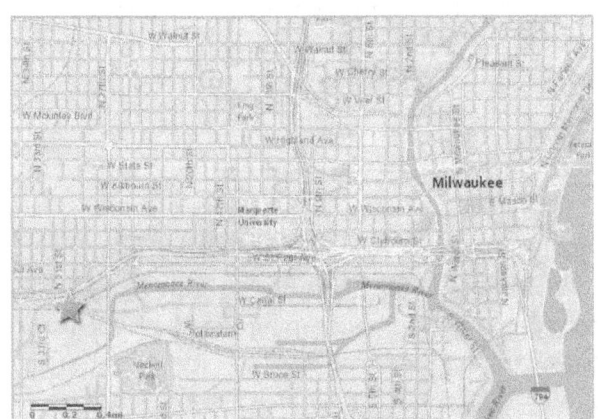

Project type:	Office and industrial development; public park; brownfield redevelopment
Green infrastructure practices:	Comprehensive site planning; stormwater treatment train, including infiltration, settling, and detention; and subsurface treatment
Completion date:	Land improvements and stormwater facility completed in 2005; private development continues as of 2015

The Menomonee Valley Industrial Center (MVIC) sits on a redeveloped former brownfield site with an industrial past dating to the late 19th century. The redevelopment involved remediating contamination and returning the vacant site to productive use as an economic engine that generates more than $1 million a year in property tax revenues and employs more than 1,400 people in a new industrial center. A centralized green infrastructure stormwater management system achieves both water quality and volume reduction objectives for current and future development while giving the community a new recreational park that provides a new access point to the Menomonee River.

Exhibit 26. The Milwaukee Road Shops were abandoned in 1985, leaving a contaminated site and an eyesore.

The City of Milwaukee

A. SITE CONTEXT

MVIC is in the Menomonee River Valley region of Milwaukee. The area was once covered with an expansive wild rice marsh that sustained native tribes for centuries. The Menomonee River is 75 miles long from its headwaters to Lake Michigan, and its watershed is approximately 140 square miles of urban landscape. Much of the river was channelized in the late 1800s as the marsh was filled to create land suitable for industrial activities along its banks.[85] A variety of contaminants impair the river's water quality, including pathogens, PCBs, phosphorus, and metals. The MVIC site has been a major source of such pollution to the river.

Industrialization of the area proceeded rapidly after the incorporation of Milwaukee in 1846. Industrial processing, manufacturing, stockyards, rendering plants, and shipping came to dominate the area.[86] Beginning in 1879 and continuing for more than 100 years, the MVIC site was home to the former Milwaukee Road Shops, which built and serviced railroad cars and locomotives. The facility closed in 1985, and the area was abandoned, leaving behind vacant, dilapidated buildings and a host of contamination issues in the underlying soils and groundwater (Exhibit 26).[87] For years, unknown remediation costs discouraged private-sector redevelopment of the

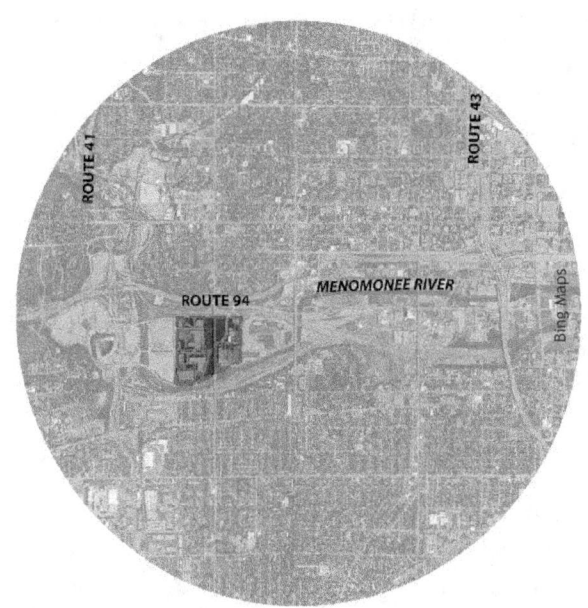

Exhibit 27. The MVIC site is located along the Menomonee River in a compactly developed part of the city.

site. In 2003, the Redevelopment Authority of the City of Milwaukee acquired the property and prepared a comprehensive master plan for its redevelopment.[88] The plan called for the demolition of existing infrastructure, remediation of polluted areas, and the redevelopment of the site into a new industrial center and recreational park.

B. PLANNING AND REGULATORY CONTEXT

The city of Milwaukee recognized the MVIC site as an important part of any economic planning in the metropolitan area due to its central location, proximity to existing transportation infrastructure, and history as the region's employment base. Employment in the Menomonee Valley had fallen from about 50,000 people in the 1920s to just over 7,000 by 1997.[89]

In 1998, the city prepared a plan for the redevelopment of the Menomonee Valley that reflected the community's goals to attract and retain industry; improve transit, biking, and walking connections in the area; and add new green space for improved aesthetics and flood control.[90]

[85] Menomonee Valley Benchmarking Initiative. *2013 Menomonee Valley State of the Valley Report*. 2014. http://www.renewthevalley.org/documents/160-resource-library.

[86] City of Milwaukee. *Menomonee Valley 2.0 Comprehensive Area Plan*. 2015. http://www.planthevalley.org/uploads/1/9/0/4/19044935/menomonee_valley_plan_5-22-15_draft.pdf.

[87] Misky, David P., and Cynthia L. Nemke. "From Blighted to Beautiful." *Government Engineering*. May-June 2010.

[88] De Sousa, Christopher. *Milwaukee's Menomonee Valley: A Sustainable Re-Industrialization Best Practice*. University of Illinois at Chicago. 2012. http://www.uic.edu/orgs/brownfields/research-results/documents/Menomonee Valley.pdf.

[89] Menomonee Valley Benchmarking Initiative *op. cit.*

[90] Rouse, David, and Ignacio Bunster-Ossa. "Menomonee Valley Park and Redevelopment, Milwaukee." In *Green Infrastructure: A Landscape Approach*. American Planning Association. 2013.

In 2002, a national design competition for the MVIC site led to implementation of a plan that included industrial development adjacent to a community park. "Stormwater Park" incorporates a centralized stormwater management facility that infiltrates, detains, and treats stormwater runoff from approximately 70 percent of the 100-acre watershed prior to discharge into the Menomonee River.

C. DESIGN AND PERFORMANCE

The project had to comply with surface water and stormwater regulations established in 2001 by the Milwaukee Metropolitan Sewer District, the City of Milwaukee Stormwater Management Regulations, and the Sustainable Design Guidelines for the Menomonee River Valley.[91]

A centralized stormwater facility meets or exceeds water quality requirements for current and future development in MVIC through a restored landscape that mimics natural river hydrology and improves water quality while reducing peak stormwater flows. Runoff is first collected and piped to a series of small ponds that allow large particulates to settle before it spreads out across a shallow wetland. Some of the stormwater is evapotranspired through meadow plants while the rest filters through the soil into a 2-foot layer of crushed lime-based concrete, recycled from a nearby highway interchange project. This subsurface infiltration area removes pollutants, provides additional storage capacity for larger storm events, and allows surface water to remain shallow enough to support the growth of wetland plants. A clay liner protects ground water. Stormwater then flows through an outlet and a subsurface treatment system to a constructed, forested wetland. Any overflow from the wetland crosses a stone river terrace leading to the Menomonee River, allowing people direct access to the river's edge for the first time in decades (Exhibit 28).[92]

Modeling indicates that the treatment system reduced total suspended solids by 80 percent, total phosphorus by 66 percent, total Kjeldahl

The City of Milwaukee

Exhibit 28. The MVIC Stormwater Park manages stormwater runoff while providing a new public space with pedestrian and bicycle trails.

nitrogen[93] by 62 percent, petroleum hydrocarbons by 76 percent, zinc by 62 percent, copper by 62 percent, and lead by 76 percent. The project is designed to contain the 2-year storm event within the constructed stormwater facilities and the 100-year storm within the entire green space of Stormwater Park. To meet the city's requirement to control peak flows during a 100-year storm event, the stormwater management facilities had to be created by reshaping and filling the flood plain. An innovative agreement with the Wisconsin Department of Transportation allowed the MVIC developers to use fill and recycled concrete from a local highway reconstruction project. The site reused 700,000 cubic yards of material that otherwise would have been deposited in a landfill, saving both projects a considerable amount of money.[94]

As of 2015, Stormwater Park has successfully handled several significant storm events and

[91] Menomonee Valley Partners. *Sustainable Design Guidelines for the Menomonee River Valley.* 2006. http://www.renew thevalley.org/media/mediafile_attachments/06/46-guidelines.

[92] Rouse and Bunster-Ossa *op. cit.*

[93] Total Kjeldahl nitrogen is the sum of free ammonia and organic nitrogen compounds.

[94] Misky and Nemke *op. cit.*

continues to collect, hold, and filter rain water as expected. Maintenance primarily involves removing invasive species from the now-established native vegetation planted on the site. In addition, MVIC maintenance staff occasionally check to see if any sediment needs to be removed from the outlet channel to maintain proper flow.[95]

D.   COSTS AND FUNDING

The stormwater costs for the MVIC development totaled $1.6 million with funding provided by federal, state, local, and various other grants. Operations and maintenance cost an estimated $100,000 per year.[96] Fees assessed to city property owners based on the amount of impervious surface area on their properties help cover these costs. Each MVIC property owner pays 40 percent of the stormwater fee to the city and the remaining 60 percent to the Redevelopment Authority of the City of Milwaukee, which uses these funds to manage Stormwater Park. This arrangement provides a dedicated source of revenue for the park, reducing the potential for future stormwater fee increases to cover maintenance or enhancements.[97]

Exhibit 29. For the first time in decades, people can access the Menomonee River from the MVIC site.

E.   BENEFITS

The MVIC project revitalized a vacant, blighted area into a thriving industrial park with public green space, including sports fields and a canoe launch. More than 60 acres of new park and open space provides public access to the Menomonee River along a stretch of the river that had been inaccessible for more than 50 years (Exhibit 29). A pedestrian and bicycle bridge across the river links the site to 7 miles of regional trails connecting the site to greater Milwaukee.[98] More than 22,000 people use the park annually, spanning all four seasons.[99]

Stormwater Park reduces pollution flowing to the Menomonee River and controls 100-year flood volumes from a 100-acre area.[100] Its ability to treat the entire development at levels exceeding regulatory requirements is an attractive incentive for businesses to locate there because it eliminates the need for prospective developers to construct individual on-lot stormwater systems that would reduce the amount of developable land.[101] Clustering development and designing shared stormwater facilities increased the developable area by 10 to 12 percent over

[95] Personal communication with David Misky, Assistant Executive Director, Redevelopment Authority of the City of Milwaukee, on Jan. 7, 2016.
[96] Cost data provided by CH2M Hill.
[97] Personal communication with David Misky op. cit.
[98] Landscape Architecture Foundation (LAF). "Menomonee Valley Redevelopment and Community Park." http://landscapeperformance.org/case-study-briefs/menomonee-valley-redevelopment-and-community-park#/overview. Accessed June 24, 2015.

[99] Urban Ecology Center. Menomonee Valley Research and Citizen Science 2014 Review. 2014. https://www.scribd.com/fullscreen/271050135?access_key=key-2qX62PoQbrVXTeogPQuH&allow_share=false&escape=false&show_recommendations=false&view_mode=scroll.
[100] LAF. "Menomonee Valley Redevelopment and Community Park." op. cit.
[101] Misky and Nemke op. cit.

conventional practices such as detention ponds for flood control.[102]

An analysis estimated that the site's aesthetic, ecologic, and recreational value increased by $120 million after redevelopment.[103] A 140-acre contaminated site was cleaned up, improving public health and returning unproductive land to productive use.[104] Over 3,000 feet of riverbank stabilization and more than 500 new native trees improved the river's water quality, wildlife habitat, and area aesthetics.[105] An ongoing citizen science monitoring program in

Stormwater Park, the adjacent Three Bridges Park, and the Hank Aaron State Trail showed that four bat species, 24 bird species, snakes, foxes, coyotes, mink, and other mammals use the area.[106]

Economic benefits have been substantial as well. As of 2015, the MVIC has 10 firms with more than 1,400 employees.[107] In addition, property values at the site increased 1,400 percent between 2002 and 2009, adding more than $1 million per year to city property tax revenues.[108] As of 2015, only 5.3 out of 60 developable acres remain.[109]

F. LESSONS LEARNED

- Incorporating community outreach early in the design process was instrumental to the project's success. The outreach attracted volunteers from local schools, businesses, and neighborhood associations who regularly plant new trees and shrubs, remove invasive species, and pick up trash.[110] As a result, the community has a sense of stewardship in Stormwater Park and its green infrastructure components.

- Early recognition of opportunities to coordinate with nearby construction projects can lead to sharing services and materials that can save a lot of money. MVIC developers reused 700,000 cubic yards of fill and recycled concrete from a local highway reconstruction project.

- Including public benefits in industrial redevelopment projects can help generate long-term community support. Stormwater

Wenk Associates

Exhibit 30. Trails through the park let visitors observe stormwater management in action.

Park provides a new access point to the river and connects the area to the regional bike and pedestrian trail system, creating an industrial area that is an economic engine for the community and an important public amenity.

[102] LAF. "Menomonee Valley Redevelopment and Community Park." op. cit.

[103] Brownfield Renewal. "Menomonee Valley Industrial Center." http://www.brownfieldrenewal.com/renewal-award-project-environmental_impact-menomonee_valley_industrial_center-8.html. Accessed Jun. 24, 2015.

[104] LAF. "Menomonee Valley Redevelopment and Community Park." op. cit.

[105] Ibid.

[106] Urban Ecology Center op. cit.

[107] Daykin, Tom. "City to Study Expansion of Menomonee Valley Industrial Center." Journal Sentinel. Feb. 19, 2015. http://www.jsonline.com/business/city-to-study-expansion-of-menomonee-valley-industrial-center-b99448425z1-292737391.html.

[108] LAF. "Menomonee Valley Redevelopment and Community Park." op. cit.

[109] Menomonee Valley Partners, Inc. "Available Properties." http://www.renewthevalley.org/categories/11-development/documents/29-available-properties. Accessed Jul. 2, 2014.

[110] NALGEP. "Spotlight on Milwaukee: Industrial Center and Community Park is Model of Sustainable Redevelopment." Mar. 19, 2014. http://www.nalgep.org/news/19/15/Spotlight-on-Milwaukee-Industrial-Center-and-Community-Park-is-Model-of-Sustainable-Redevelopment.html.

G. PROJECT TEAM

- **Owner and developer:** The Redevelopment Authority of the City of Milwaukee
- **Engineering and environmental remediation:** Milwaukee Transportation Partners (CH2M Hill and HNTB)[111]
- **Lead planner and landscape architect:** Wenk Associates[112]

[111] Rouse and Bunster-Ossa *op. cit.*

[112] Ibid.

VIII. UPTOWN NORMAL CIRCLE
NORMAL, ILLINOIS

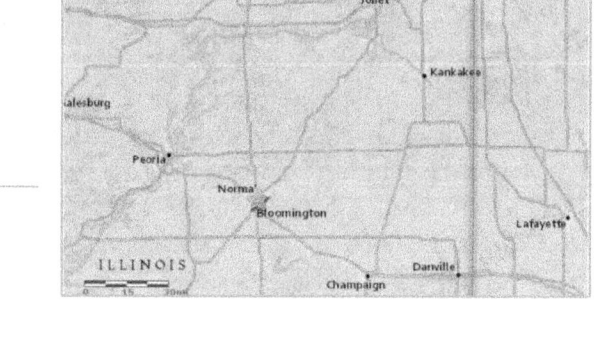

An innovative roundabout calms traffic and creates a new public gathering space in a stormwater park, helping to revitalize a central business district.

Project type:	Transportation; public plaza
Green infrastructure practices:	Underground cistern, filtration bogs, and structural cell and conventional tree planters
Completion date:	2010

The Uptown Normal Circle is a $15.5 million redevelopment project that includes an innovative stormwater management system, new streets, and renovated streetscapes for the core of the six-block Uptown area. A new roundabout calms traffic at a previously chaotic three-way intersection while creating a new public gathering place within a stormwater park. The project's main goal was to catalyze revitalization of the central business district while showcasing sustainability practices. It manages stormwater from nearly 3 acres of impervious cover in the central business district and is a valued public amenity that has attracted new private investment, helped spur continued downtown redevelopment, and received national recognition.[113]

A. SITE CONTEXT

Normal is a town of just over 50,000 people in central Illinois. It was laid out in 1865 at the confluence of the Chicago and Alton Railroad and the Illinois Central Railroad. The town population has grown slowly but steadily, but Normal's town center began to decline as early as the 1950s as downtown businesses started closing.[114] By the late 1990s, Normal suffered from storefront

---

[113] Among other awards, this project received EPA's National Award for Smart Growth Achievement. See: EPA. *2011 National Award for Smart Growth Achievement Booklet.* https://www.epa.gov/smartgrowth/2011-national-award-smart-growth-achievement-booklet.

[114] Gorsche, Jennifer K. "Circular Logic Reshapes Downtown Normal." *The Architect's Newspaper.* Aug. 16, 2010. http://archpaper.com/news/articles.asp?id=4768#.Va5Bik0w_Gh.

vacancies, declining property values, and nearly 100-year-old public infrastructure in dire need of updating.[115]

B.   PLANNING AND REGULATORY CONTEXT

In 1999, Normal embarked on an ambitious redevelopment project to catalyze revitalization in the town center and help reverse the downward trajectory. The town developed the Downtown Normal Redevelopment Plan[116] after community input from more than 70 public meetings. In 2001, the town council adopted the plan, which incorporates environmental sustainability as a strategy to help boost the economy. It included one of the first ordinances in the country to require buildings over 7,500 square feet to meet green building standards, plans for a multimodal transportation center that links local and regional transit, and streetscape improvements to create a more walkable town center.[117]

The Uptown Normal Circle was a key recommendation of the redevelopment plan. The project focused on resolving long-standing design problems arising from an awkward intersection of three streets that divided the central business district. Designers proposed a new traffic roundabout to slow traffic, improve pedestrian

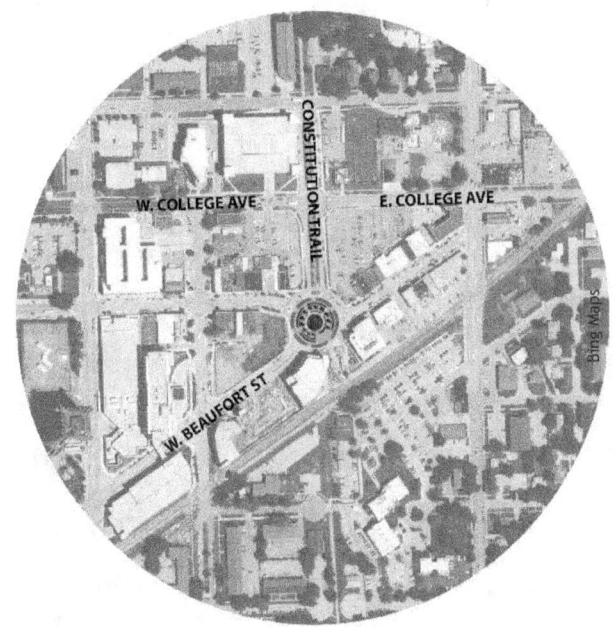

Exhibit 31. The Uptown Normal Circle sits at a formerly chaotic three-way intersection.

safety, and use green infrastructure to manage stormwater, while providing an interactive green space for the community to enjoy.[118]

The circle was designed to convey a 10-year storm event, meeting state and local standards for stormwater management. The project was not subject to any water quality treatment standards, so all of the water quality enhancement elements of the project were voluntary.

C.   DESIGN AND PERFORMANCE

Green infrastructure is a key feature of the circle, which captures, stores, cleans, and recycles water from several streets surrounding the circle. A 75,000-gallon underground cistern, created from an abandoned storm sewer line, provides storage space for stormwater runoff. This recycled water helps irrigate the six-block core of the central business district. The water circulates by gravity through a series of terraced

filtration bogs in the circle.[119] From the bogs, cleansed water flows first into a collection pool and then into a secondary underground reservoir where it is treated by ultraviolet light to destroy microorganisms. From the secondary reservoir, water is pumped through a shallow, stream-like fountain that people can dip their feet in to cool

[115] Town of Normal. "History of Redevelopment." https://www.normal.org/index.aspx?NID=832. Accessed Jul. 20, 2015.
[116] The Normal Town Council voted to change the name of "Downtown Normal" to "Uptown Normal" in 2006.
[117] Town of Normal, "History of Redevelopment" op. cit.

[118] Gray, Rob. "Sustainability as Catalyst: Uptown Normal Circle." APWA Reporter. Apr. 2011: 66-70. http://www.apwa.net/Resources/Reporter/Articles/2011/4/Sustainability-as-catalyst-Uptown-Normal-Circle.
[119] Ibid.

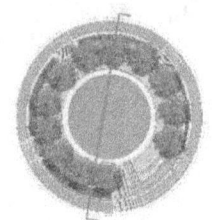

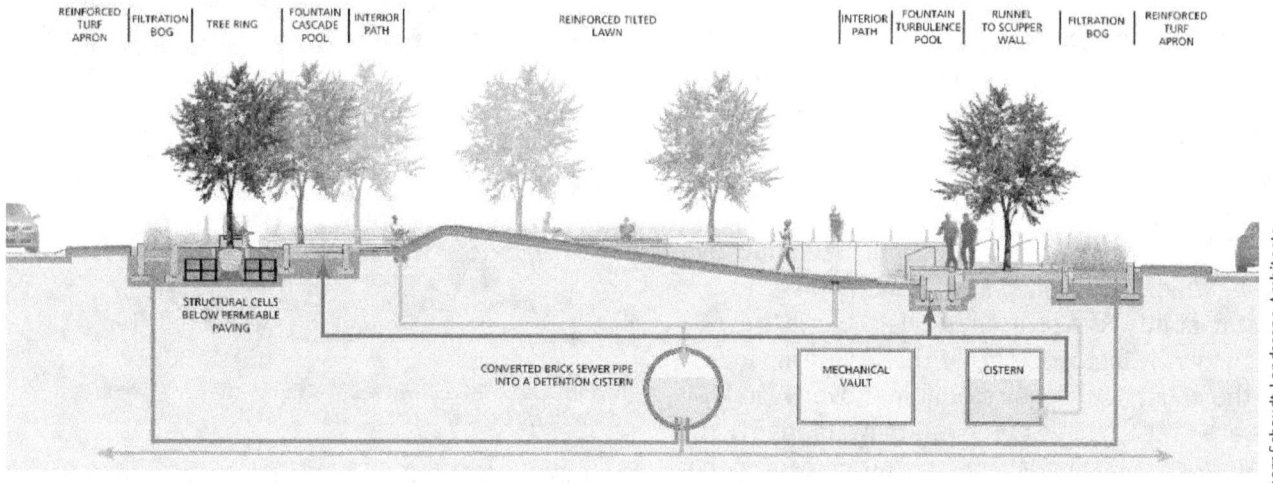

DETENTION CISTERN SUPPLY FROM STORMWATER
DISPLAY FOUNTAIN SUPPLY
DISPLAY FOUNTAIN RETURN
FILTRATION BOG SUPPLY
IRRIGATION SUPPLY

REINFORCED TURF APRON | FILTRATION BOG | TREE RING | FOUNTAIN CASCADE POOL | INTERIOR PATH | REINFORCED TILTED LAWN | INTERIOR PATH | FOUNTAIN TURBULENCE POOL | RUNNEL TO SCUPPER WALL | FILTRATION BOG | REINFORCED TURF APRON

STRUCTURAL CELLS BELOW PERMEABLE PAVING

CONVERTED BRICK SEWER PIPE INTO A DETENTION CISTERN

MECHANICAL VAULT

CISTERN

Hoerr Schaudt Landscape Architects

Exhibit 32. A cross section of the circle shows the flow of water through the system.

off and that creates sound to mask the nearby traffic noise (Exhibit 32).[120]

The project also incorporates infiltration planters and underground structural cells for tree plantings in a ring around the circle and along the nearby sidewalks that help prevent soil compaction and retain absorptive capacity. Although water quality treatment was not required, the system is estimated to remove an estimated 91 percent of total suspended solids, 79 percent of total phosphorus, and 64 percent of total nitrogen from stormwater.[121]

D.   COSTS AND FUNDING

The entire redevelopment for the circle and surrounding streets, including utilities, roads, streetscape improvements, and landscaping, totaled approximately $15.5 million, with stormwater costs of $1.3 million, about half of which was for aesthetic features and half of which was for functional components.[122] The town recognized early on that the downtown redevelopment plan would have a greater chance of implementation with a dedicated source of local funding. New revenue sources dedicated solely to the redevelopment effort included a sales tax, a hotel/motel tax, a food and beverage tax, and establishment of a tax-increment financing district that allowed future increases in property taxes in the area to be directed to area improvements.[123] Other sources of funding for the overall redevelopment plan included municipal bonds and grants from the Federal Transit Administration and the Illinois Department of Commerce and Economic Opportunity.[124] A U.S. Department of Transportation TIGER grant helped fund the multimodal transportation center. In addition, in

[120] Town of Normal. *The Uptown Normal Circle*. Undated. http://www.normal.org/DocumentCenter/View/4409.
[121] Landscape Architecture Foundation (LAF). "Uptown Normal Circle and Streetscape." http://landscape

performance.org/case-study-briefs/uptown-normal-circle-and-streetscape. Accessed Jul. 20, 2015.
[122] Cost data provided by the town of Normal.
[123] Town of Normal, "History of Redevelopment" *op. cit.*
[124] Gray *op. cit.*

2006, the town enacted a stormwater utility fee that generates about $1.8 million per year for maintenance and construction of new stormwater management practices.[125]

E. BENEFITS

The site captures 1.4 million gallons of stormwater annually, reducing the burden on the municipal stormwater system.[126] More than 100 new trees sequester nearly 11,000 pounds of carbon annually and reduce ambient temperatures. Because of the use of structural planting cells, the trees have an expected lifespan triple that of conventionally planted street trees.[127]

In addition to environmental benefits, the project created a place where the community gathers for special events and daily use. More people now walk and bike to the Uptown District, and the project has attracted new businesses and people. After completion of the redevelopment plan, including the Uptown Circle, reconstruction of Constitution Boulevard, and construction of the transportation center, private businesses invested $160 million in the Uptown District, including the construction of a

Exhibit 33. The Uptown Normal Circle serves multiple purposes in a small space previously used solely for traffic.

new hotel and conference center. Property values went up 16 percent, and retail sales grew 46 percent.[128] At least four organizations chose to hold conferences in Normal that featured the completed circle, bringing nearly $700,000 in tourism dollars to the city.[129]

F. LESSONS LEARNED

- Environmental sustainability initiatives can help generate economic development. The Downtown Normal Redevelopment Plan's focus on sustainability, including the Uptown Normal Circle, has garnered nationwide recognition that has helped attract conferences and private investment.

- Identifying a locally generated funding source for green infrastructure projects like Normal's stormwater utility fee can help ensure both their implementation and long-term success. While federal and state funds can help significantly with capital costs, long-term maintenance generally is a local responsibility. Maintenance is particularly

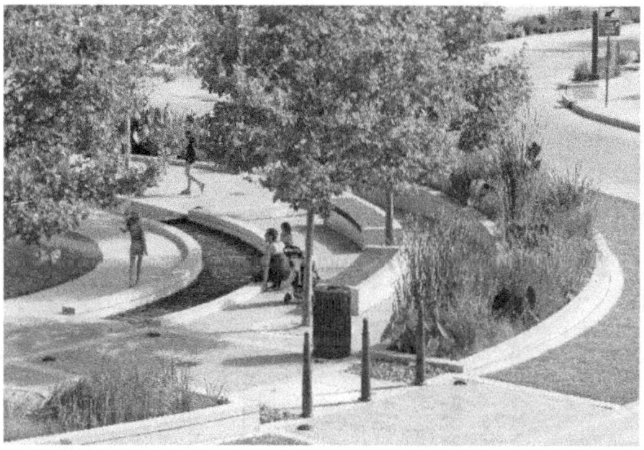

Exhibit 34. Stormwater features in the Uptown Normal Circle manage runoff while creating a beautiful park for the public to enjoy.

[125] Aldrich, Wayne. "Storm Water Utility Fees." *WTVP At Issue*. Episode #2728. May 14, 2015. http://www.wtvp.org/programming/ai/2-2728.asp.
[126] LAF. "Uptown Normal Circle and Streetscape." *op. cit.*
[127] Ibid.

[128] Smart Growth America and National Complete Streets Coalition. *Safer Streets, Stronger Economies: Complete Streets Project Outcomes from Across the Country*. 2015. http://www.smartgrowthamerica.org/research/safer-streets-stronger-economies.
[129] LAF. "Uptown Normal Circle and Streetscape." *op. cit.*

important for green infrastructure projects because they are so visible to the community.

- Community leaders' support can be critical to overcome public skepticism about the value of spending public dollars on aesthetic improvements. Educating the public about the multiple environmental, economic, and social benefits was important to generate community support in Normal for new types of infrastructure investments.

G. PROJECT TEAM

- **Owner and developer:** Town of Normal, Illinois
- **Engineers:** Clark Dietz Incorporated (roadway design) and Farnsworth Group (underground infrastructure design)
- **Landscape architect:** Hoerr Schaudt Landscape Architects
- **Master planners:** Farr Associates

Exhibit 35. The Uptown Normal Circle attracts residents and visitors to the business core by providing a gathering place.

IX. THE METRO GREEN LINE
ST. PAUL, MINNESOTA

Green infrastructure along a new light rail corridor provides stormwater management for the largest public works project in Minnesota history.

Project type:	Transportation; brownfield redevelopment
Green infrastructure practices:	Integrated tree trench system with structural soil, stormwater planters, rain gardens, and infiltration trenches
Completion date:	2012; rail service began in 2014

The Metro Green Line (formerly the Central Corridor Light Rail Transit project) covers 11 miles, connecting the major downtown areas of the Minnesota state capital in St. Paul with Minneapolis. This case study focuses on the 7.5-mile section of the project that is in St. Paul and under the jurisdiction of the Capitol Region Watershed District (CRWD). CRWD regulations required the Green Line project to include stormwater quality improvements, preferably infiltration, wherever feasible. However, an extensive system of underground utilities provided little space. To address this challenge, the Metropolitan Council, the regional planning agency serving St. Paul and Minneapolis' seven-

county metropolitan area, developed an innovative stormwater management system that includes an integrated tree trench system. In addition, CRWD and the city of St. Paul augmented stormwater management in the corridor with stormwater planters, rain gardens, and infiltration trenches on side streets. Green infrastructure was a cost-effective way to comply with stormwater regulations while providing additional benefits that helped gain public support from the surrounding community for the project. A new canopy of more than 1,250 trees along the heavily developed route will provide shade, beautify the area, and improve air and water quality.

A. SITE CONTEXT

The city of St. Paul lies mostly on the north bank of the Mississippi River adjacent to Minneapolis.

Most of the light rail route follows University Avenue, one of the oldest streets in the metropolitan area. It was served by streetcars from 1890 until 1954 and developed with a mix of manufacturing, retail, hospitals, offices, entertainment venues, and housing.[130] The corridor is racially and ethnically diverse and has higher levels of poverty than the surrounding metropolitan region.[131]

Before construction, the 120-foot-wide right-of-way for the rail line was mostly impervious, including a four-lane road from which all stormwater runoff flowed untreated directly to the Mississippi River through numerous outfalls, carrying sediment and pollution. The relatively narrow project space, the city's desire to accommodate compact development near rail stations, and a prohibition on infiltration into the road subbase limited the types of green infrastructure that would be suitable. Contaminated soils and shallow groundwater limited green infrastructure to approximately 50 percent of the St. Paul segment of the project.

B.   PLANNING AND REGULATORY CONTEXT

In 2003, CRWD began developing stormwater management guidelines for development and redevelopment in the area. However, two years after issuing the guidelines, the district found that many developers were not installing any stormwater management controls, so CRWD decided to develop formal regulatory rules.[132]

In 2006, CRWD issued water quality and stormwater management rules for projects disturbing more than 1 acre of land. The rules require reducing pollution flowing to lakes, wetlands, and the Mississippi River by meeting standards for runoff rate, volume reduction, and water quality. A volume equal to 1 inch of rainfall from impervious surfaces on the site must be retained on-site, and best management practices must remove 90 percent of total suspended solids from the runoff generated by a 2.5-inch rainfall event.[133]

During the rulemaking process, commenters were concerned with the cost of compliance for major public transportation projects, particularly linear projects that generally have space constraints, extensive utilities, and a high percentage of impervious cover. These concerns led to a cost cap for linear projects that limits costs for complying with stormwater regulations to $30,000 per acre of new or reconstructed impervious surface, an amount that is set annually by the CRWD Board.[134]

C.   DESIGN AND PERFORMANCE

CWRD and its partners convened a stormwater design workshop in June 2009 with more than 50 participants. The workshop led to development of conceptual plans for stormwater management along the Green Line that would meet the CRWD stormwater standards cost-effectively while achieving other community goals such as cleaner air, more green space, and improved aesthetics along the corridor.[135]

The centerpiece of the design is an integrated tree trench system that accommodates the site's limitations. It can infiltrate runoff from the roads and rail line while safely supporting traffic

[130] Isaacs, Aaron. "Rail Returns to the Central Corridor." MetroTransit blog. Jun. 11, 2014. http://www.metrotransit.org/rail-returns-to-the-central-corridor. Accessed Jan. 15, 2015.

[131] PolicyLink, TakeAction Minnesota, and ISAIAH. *Healthy Corridor for All: A Community Health Impact Assessment of Transit-Oriented Development Policy in Saint Paul, Minnesota.* 2011. http://isaiahmn.org/2012/01/healthy-corridor-for-all.

[132] CRWD. "Watershed Rules." http://www.capitolregionwd.org/permits/watershed-rules. Accessed Jan. 15, 2015.
[133] Ibid.
[134] Ibid.
[135] Eleria, Anna, and Forrest Kelley. "Green Infrastructure for the Central Corridor Light Rail Transit Project." 2013 International Low Impact Development Symposium. Aug. 18-21, 2013. http://assets.conferencespot.org/fileserver/file/34648/filename/a621_1.pdf.

loads and protecting tree roots. A PVC barrier keeps infiltrated water away from the road subbase. To avoid existing utilities and limit the impacts on the existing road, the tree trench system is located under curbing, sidewalks, and boulevards along 5.2 miles of University Avenue where the soil is suitable for infiltration (Exhibit 36).

The system includes permeable pavers and "structural soil," which is gravel with a specialized soil coating similar in size and function to conventional load-bearing subbase materials. It can support heavy loads of foot and vehicle traffic along the corridor, and tree roots can grow through it. Rainfall on the sidewalks infiltrates through permeable pavers and structural soils, supplying water and air to the tree roots, while catch basins and a perforated pipe direct runoff from the road to infiltration chambers.

The system supports 1,250 new trees along the corridor and reduces runoff to the maximum extent practicable in the limited available space in compliance with the CRWD requirements. CRWD and the city of St. Paul installed additional green infrastructure practices along the corridor and adjacent streets, including stormwater planters, rain gardens, and infiltration trenches.[136] Maintenance of the green infrastructure practices on side streets with high pedestrian traffic involves biweekly trash and debris removal and annual weeding and plant replacement.[137]

Pre-construction estimates for the performance of the integrated tree trench system and other green infrastructure practices were that they would reduce stormwater volume by 50 percent, phosphorus loading by 85 pounds, and sediment loading by 20,000 pounds per year, helping to improve water quality in the Mississippi River.[138] Testing after construction showed that the tree trench system is exceeding its performance

targets.[139] Monitoring has been continuing since 2013, and CWRD plans to issue a report on performance and pollutant removal effectiveness in 2016.[140] Although the project does not strictly meet volume reduction standards, it complies with the stormwater regulations because it exceeded the cost cap set for linear projects.

Exhibit 36. A tree trench system along University Avenue allows room for roots to grow in a highly trafficked area.

Additional features of the light rail project include a pedestrian mall, improved sidewalks and crosswalks, bike racks, planters, benches, permeable pavers, and LED lighting. These features work together with the green infrastructure to make a pleasant and safe environment for people walking and biking to the light rail stations.

[136] Minnesota Pollution Control Agency. "Case Studies for Tree Trenches and Tree Boxes." http://stormwater.pca.state.mn.us/index.php/Case_studies_for_tree_trenches_and_tree_boxes. Accessed Jan. 16, 2015.
[137] Ibid.

[138] CRWD. *2014 MAWD Project & Program of the Year.* 2014. http://www.capitolregionwd.org/wp-content/uploads/2014/10/Capitol-Region-project-of-the-year-final-nomination.pdf.
[139] Eleria and Kelley *op. cit.*
[140] Personal communication with Mark Doneux, Administrator, CRWD, on Nov. 25, 2015.

D. COSTS AND FUNDING

The total project cost for the Green Line was nearly $1 billion, the largest completed public works project in Minnesota history.[141] Multiple sources provided financial support (Exhibit 37).

Stormwater costs for the Green Line totaled $5,114,865, or 0.5 percent of the total project costs. Funding for the stormwater costs came from a Minnesota Clean Water Legacy Fund grant for $665,000, and contributions from CWRD, the Metropolitan Council, and the city of St. Paul.[142]

FUNDING SOURCE	PERCENTAGE
Federal Government	50
Counties Transit Improvement Board	30
Minnesota state	9
Ramsey County	7
Hennepin County	3
Metropolitan Council	1
City of St. Paul and Central Corridor Funders Collaborative	<1

Exhibit 37. Total project funding.

Source: Metropolitan Council. "Project Funding." http://www.metro council.org/Transportation/Projects/Current-Projects/Central-Corridor/Grants-Funding-(CCLRT).aspx. Accessed Jan. 15, 2015.

E. BENEFITS

The green infrastructure practices developed for the Green Line allowed the city to meet its goals of improving regional transportation options, facilitating compact redevelopment along the new transit corridor, and improving water quality in the Mississippi River. The Green Line is exceeding ridership expectations, with 45,644 daily riders as of September 2014, more than originally predicted to occur by 2030.[143]

A new tree canopy in this densely developed area with limited green space helps capture stormwater, improve air quality, reduce temperatures during hot weather, and beautify the neighborhood. Because of the trench system, the 1,250 new trees are much more likely to survive in the harsh urban environment and will require less irrigation than other street trees, saving maintenance costs. Because melting snow will infiltrate rather than refreezing, the permeable sidewalks will require less salt in the winter to maintain a safe walking surface, saving money and reducing salt going to the Mississippi River. Interpretive signage along the Green Line in English, Spanish, and Hmong helps educate rail users about the need for stormwater pollution controls and how the green infrastructure

Capitol Region Watershed District

Exhibit 38. Signs help passersby understand the importance and function of rain gardens and other green infrastructure.

---

[141] Metropolitan Council. "Metro Green Line Opens On Time and On Budget." Jun. 14, 2104. http://www.metrocouncil.org/News-Events/Transportation/News-Articles/METRO-Green-Line-opens-on-time-and-on-budget.aspx.

[142] Buranen, Margaret. "Green Infrastructure Makes Sen$e in the Twin Cities." Stormwater. Jan./Feb. 2013. pp. 8-11.

http://digital.stormh20.com/publication/index.php?p=11&i=138017&ver=swf&pp=2&zoom=0.

[143] Personal communication with Anna Eleria, CRWD, on Nov. 25, 2015.

installations improve water quality (Exhibit 38).[144]

Construction of the Green Line spurred more than $2.5 billion in private redevelopment along the corridor even before starting operation in 2014 (Exhibit 39).[145] In spite of disruptions due to construction of the Green Line, there was a net gain of 13 businesses directly on the line during the construction period between February 2011 and June 2014. In addition, 4,459 market-rate and 2,375 new or preserved long-term affordable housing units were created during this period.[146]

Exhibit 39. New businesses along the Green Line can benefit from the increased foot traffic it brings to neighborhoods.

F.   LESSONS LEARNED

- Implementing green infrastructure in a major public project can encourage wider use of green infrastructure by demonstrating its benefits. The city of St. Paul and CRWD installed additional green infrastructure on adjacent streets along the Green Line corridor, and the entire area serves as a demonstration project for other developments in the city.

- Screening for soil contamination that could limit the ability to infiltrate stormwater should occur before project design to identify areas suitable for green infrastructure. Designers had to reconfigure the project after initial planning, reducing the expected benefits.

- Green infrastructure can be feasible on highly constrained sites without space for conventional stormwater management. The Green Line could not have accommodated conventional stormwater treatment in the narrow right of way available due to existing infrastructure.

G.   PROJECT TEAM

- **Owner and developer:** The Metropolitan Council

- **Engineers:** AECOM, Kimley-Horn and Associates, Inc. and HZ United[147]

---

[144] CRWD. "Green Line Green Infrastructure Practices – Water Quality." http://www.capitolregionwd.org/our-work/watershed-planning/cclrt_wq. Accessed Jan. 16, 2015.
[145] Metropolitan Council. "Metro Green Line Helps Attract at Least $2.5 Billion in Development." May 14, 2014. http://www.metrocouncil.org/News-Events/Transportation/ News-Articles/Metro-Green-Line-helps-attract-at-least-$2-5-billi.aspx. Accessed Jan. 16, 2015.
[146] Business Resources Collaborative. *Healthy Local Businesses, Healthy Communities.* 2015. http://www.funderscollaborative.org/sites/default/files/BRC_0315-1_Final_Report_10.pdf.
[147] Minnesota Pollution Control Agency *op. cit.*

X. Santa Fe Railyard Park and Plaza

Santa Fe, New Mexico

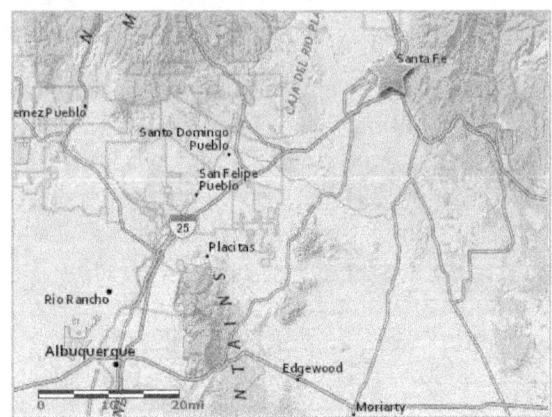

An abandoned railyard is transformed into a flourishing community activity center, including a 10-acre park that helps reduce stormwater runoff.

Project type:	Commercial development; public park and plaza; historic preservation; brownfield redevelopment
Green infrastructure practices:	Comprehensive site planning, water harvesting (cistern and water tower, swales, and two stormwater detention areas)
Completion date:	2008

The Santa Fe Railyard Park and Plaza project converted an abandoned railyard into a flourishing community activity center that incorporates green infrastructure for water conservation. The design for the site, chosen through an international design competition, incorporated significant community input. It integrates the site's historic features and open space into the fabric of downtown and adjacent neighborhoods.

The project includes a 10-acre park, a 1-acre plaza, and a 1.5-acre pedestrian walkway, which are protected under a permanent conservation easement. A water harvesting system can store 110,000 gallons of stormwater collected from impervious surfaces on the property. The project also included the restoration of the Acequia Madre, a historic irrigation canal running through the site. Water is a major aspect of this redevelopment because Santa Fe is in an arid, high-desert climate, and water conservation is a major concern.

A. SITE CONTEXT

In downtown Santa Fe, the site of the Railyard Park and Plaza was historically used for

agriculture by Native Americans and Spanish settlers. Along the site runs the Acequia Madre, one of the oldest irrigation canals in the United States dating back to the 1600s.[148]

In the 1880s, the area became the site of a railyard, with surrounding neighborhoods created to house workers and their families. By the 1940s, the area became a center of community life with gardens, swimming, and ice skating.[149] However, after the railroad suspended passenger service following World War II, the railyard and surrounding neighborhoods began to decline. By 1987, the area was declared blighted, and the city launched a master-planning process to redevelop it.[150] Due to its industrial past, the site was contaminated with lead and other metals, petroleum, and petroleum products, impeding redevelopment.[151]

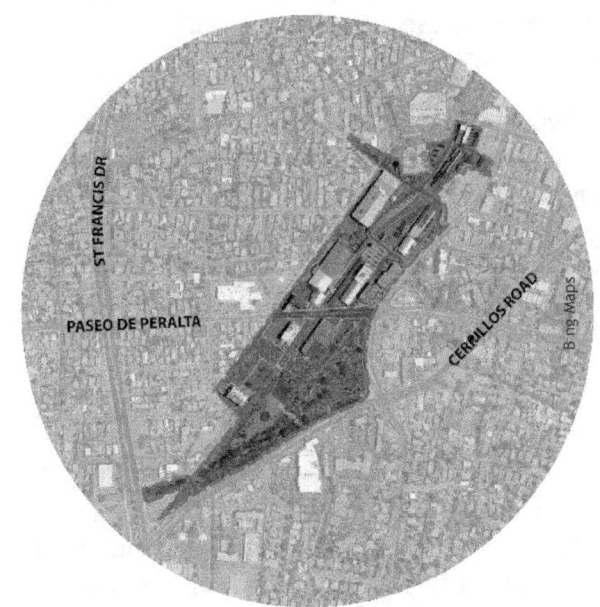

Exhibit 40. The Santa Fe Railyard Park and Plaza is a center of activity for the surrounding neighborhoods.

B. PLANNING AND REGULATORY CONTEXT

The planning process took years, as citizen activists strongly pushed to create a pedestrian-oriented area with public open space and local businesses, reestablish rail service, and preserve the character of surrounding neighborhoods. In 1995, the Trust for Public Land and the city worked together to buy the Santa Fe Railyard and redevelop the area, placing 13 of the site's 50 acres under a permanent conservation easement.[152]

Ultimately, more than 6,000 residents provided input into the railyard's redevelopment. In 2002, the city council approved the Railyard Master Plan and organized the Santa Fe Railyard Community Corporation to oversee mixed-use development for 37 acres of the site. Meanwhile, the Trust for Public Land conducted an international design competition for the railyard's public spaces. It called for a new park

and plaza integrated with the social life of the city through the protection and enhancement of historic areas.[153]

Environmental contamination at the site was cleaned up by 2006 with assistance from U.S. Environmental Protection Agency brownfields assessment grants and the New Mexico Environment Department's Voluntary Cleanup Program, setting the stage for development to begin.[154]

The city's stormwater management standards required that the stormwater runoff peak flow rate discharged from the site not exceed pre-development conditions for the 100-year, 24-hour storm event. The city's landscape and site design standard required the use of water harvesting and encouraged developing and using

[148] Crawford, Stanley. "A Central Park for Santa Fe." *Land + People.* Spring/Summer 2009. https://www.tpl.org/magazine/central-park-santa-fe%C2%97landpeople.
[149] The Santa Fe Railyard Community Corporation. "Railyard History." http://www.railyardsantafe.com/history. Accessed Aug. 11, 2015.
[150] Shibley, Robert, Brandy H.M. Brooks, Jay Farbstein, and Richard Wener. *Partnering Strategies for the Urban Edge: 2011 Rudy Bruner Award for Urban Excellence.* 2011. http://www.brunerfoundation.org/rba/pdfs/2011/2011Book.pdf.

[151] New Mexico Environment Department. "Brownfields Success Stories." https://www.env.nm.gov/gwb/NMED-GWQB-BrownfieldsSuccessStories.htm. Accessed Aug. 11, 2015.
[152] Shibley, Brooks, Farbstein, and Wener *op. cit.*
[153] Shibley, Brooks, Farbstein, and Wener *op. cit.*
[154] EPA. *Old Santa Fe Railyard: Back on Track to Revitalization.* 2008. nepis.epa.gov/Exe/ZyPURL.cgi?Dockey=P1007CTE.TXT.

sources of landscape irrigation water other than potable water.[155]

Designers faced several constraints in designing the stormwater management system for the site. Water rights are a key consideration in this arid region. Due to New Mexico's commitments under the Rio Grande Compact,[156] the Office of the State Engineer prohibits passive water harvesting—techniques that would detain water so that it could slowly infiltrate into the soil. Regulations allowed for active water harvesting—collection in a storage container for later use—but only for runoff within the 50-acre site and not from surrounding streets. In addition, the city did not allow designers to construct a decentralized wastewater treatment plant from which treated effluent could be used for irrigation.[157]

C. DESIGN AND PERFORMANCE

Runoff from approximately 3.7 acres of railyard buildings and impervious surfaces is collected in five 15,000-gallon underground storage tanks and a 35,000-gallon water tower for a total of 110,000 gallons of storage capacity that supplies irrigation for the park. The tower is not only a storage facility but also a landmark that contributes to the area's character.[158] Swales and stormwater detention facilities also help reduce runoff rates, slowing erosion. However, runoff volume is not reduced due to regulations prohibiting passive water harvesting that required water collected by these practices to be piped into the drainage system rather than infiltrated.

Exhibit 41. The Railyard Park incorporates extensive native plantings and naturalistic landscaping.

Extensive native plantings were incorporated into the landscape, which includes a shady riparian area, a dry gulch that fills seasonally with rain, ornamental gardens adapted for dry conditions, and bird and butterfly gardens (Exhibit 41). In addition, the area's historical agriculture is represented through community gardens, orchards, and historic Pueblo gardens designed for arid conditions.[159] The Acequia Madre Association granted permission for a diversion channel to provide additional irrigation water for the community gardens in the park, much as the Acequia Madre has been supplying water to the area for 400 years. The irrigation system, signage, and outdoor classrooms at the community gardens help the public understand the link between historical agriculture practices and the need for water conservation in a region with limited water resources.[160]

The water harvesting system has flow sensors that track the amount of rainwater collected and used and supplemental city water needed. The public can view monthly, annual, and accumulated historical data.[161]

[155] City of Santa Fe. "Article 14-8: Development and Design Standards." 2001. http://clerkshq.com/Content/Santafe-nm/books/landdevelopment/sfld_a8.htm.
[156] The Rio Grande Compact is a 1938 agreement among Colorado, New Mexico, and Texas that apportions the waters of the Rio Grande Basin.
[157] Personal communication with Frederic Schwartz, Frederic Schwartz Architects, on Apr. 19, 2011.

[158] Crawford op. cit.
[159] Railyard Stewards. "Horticulture in the Park." http://www.railyardpark.org/park-plaza/horticulture-in-the-park. Accessed Aug. 11, 2015.
[160] Schwartz op. cit.
[161] Ibid.

D.   COSTS AND FUNDING

The railyard redevelopment project cost $137 million in total, including an estimated $70 million in private investment as of 2011. The cost of the park, plaza, and walking path was $13 million, including $400,000 for planning, $1.1 million for design and engineering, $10.5 million for construction, and $1.5 million for administrative costs.[162] Stormwater costs amounted to about $2 million, or 15 percent of total project costs.[163] The Trust for Public Land raised $13 million from:

- State legislative appropriations ($3.1 million).

- Federal transportation funds ($2.4 million).
- City capital improvement bonds ($1.3 million).
- City and county gross receipts taxes ($600,000).
- Santa Fe Southern Railway ($2.3 million).
- Other private donors ($3.1 million).[164]

Volunteers from the Trust for Public Land formed a membership group, the Railyard Stewards, to help provide maintenance, program events, and advocate for the park. The Railyard Stewards work with the city of Santa Fe to encourage residents to visit the park and plaza.[165]

E.   BENEFITS

The Railyard Park and Plaza redevelopment project restored a former brownfield to the vibrant downtown community center it once was. The Railyard Plaza hosts performances and special events, while also providing regular space for food vendors and a farmers market featuring local growers and artisans. About 20 percent of the project site has been protected through a conservation easement as public open space. The Railyard Park includes an informal outdoor performance space, a children's play area, picnic areas, community gardens, and a walking and biking trail linked to a citywide trail. The New Mexico RailRunner Express commuter rail service stops at the historic Santa Fe Depot, building on the location's long history of train travel.[166]

The city's investment in the Railyard redevelopment project led to millions of dollars in private investment, including restaurants, shops, and a cinema.[167] More than 90 percent of tenants in the Railyard are local businesses and nonprofit organizations.

Exhibit 42. The farmers market at the Railyard Plaza is a popular gathering place.

An innovative water harvesting system, which works within the water rights restrictions common to arid regions, uses stormwater runoff to irrigate more than 300 new trees and several thousand drought-resistant and native plants.

---

[162] Crawford *op. cit.*
[163] Personal communication with Suby Bowden, Suby Bowden + Associates, on Apr. 28, 2011.
[164] Crawford *op. cit.*
[165] Railyard Stewards. "The Railyard Stewards." http://www.railyardpark.org. Accessed Aug. 11, 2015.

[166] Crawford *op. cit.*
[167] KRQE News. "Santa Fe Railyard is getting two new attractions." May 22, 2015. http://krqe.com/2015/05/22/santa-fe-railyard-adding-two-new-attractions.

F.   LESSONS LEARNED

- Green infrastructure can help achieve water conservation goals in arid climates by storing stormwater for irrigation. Even areas subject to water rights laws can incorporate green infrastructure into development projects.

- Maximizing developable area on a site is not always conducive to meeting a community's goals for that site. The Santa Fe Railyard project reduced the development density on the site in exchange for preserving historic places tied to the city's identity and creating a large open space that serves city residents and helps protect the environment. In return, the developers garnered more support for the project and sparked community pride in the project.

Exhibit 43. The Santa Fe Railyard builds on the city's unique assets to create a space that residents and visitors love.

G.   PROJECT TEAM

- **Owner:** The City of Santa Fe
- **Developer:** The Trust for Public Land
- **Engineering:** URS
- **Architect:** Fredric Schwartz

- **Landscape architect:** Ken Smith
- **Landscape artist:** Mary Miss
- **Landscape design:** Edith Katz

XI. STAPLETON GREENWAY PARK
DENVER, COLORADO

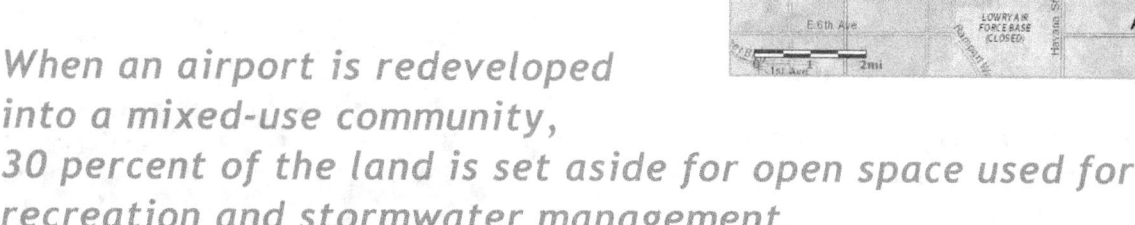

When an airport is redeveloped into a mixed-use community, 30 percent of the land is set aside for open space used for recreation and stormwater management.

Project type:	Mixed-use development; public park
Green infrastructure practices:	Vegetated swales and constructed wetland
Completion date:	2002

The Stapleton Airport redevelopment converted an obsolete airport just 6 miles from downtown Denver into a mixed-use community with single- and multifamily homes, businesses, restaurants, office space, and schools, setting aside 30 percent of the area for open space. Its public parks and greenways are a key selling point and a cherished amenity for those who live in, work in, or visit Stapleton. The developer integrated green infrastructure into the parks and landscape, creating centralized facilities that simultaneously meet water quality, flood control, and open space requirements.

A.   SITE CONTEXT

In 1989, city leaders in Denver, Colorado, decided to build a new airport. By abandoning Stapleton International Airport, they created a 4,700-acre redevelopment opportunity in an already-developed area 6 miles from downtown.[168]

Westerly Creek once flowed through the site, but construction of the airport in 1929 enclosed the creek in two underground pipes. The Westerly Creek watershed covers 18.5 square miles of mostly developed land, leaving the area subject to seasonal flooding when stormwater flows exceed the capacity of the piping system.[169] In addition, stormwater runoff from the site caused both surface and groundwater contamination.[170]

---

[168] Carder, Carol. "New Life at an Old Airport." *Progressive Engineer*. 2011. http://www.progressiveengineer.com/features/new_life_old_airport.htm.
[169] Ibid.

[170] City and County of Denver, Stapleton Redevelopment Foundation, and Citizens Advisory Board. *Stapleton Development Plan*. 1995. http://stapletonfoundation.com/wp-content/uploads/2015/05/GreenBook1995_ForWeb Viewing.pdf.

B. PLANNING AND REGULATORY PROCESS

The city and county of Denver, Denver International Airport, and citizen advisory groups began planning for redevelopment before the airport closed. A two-year community planning process developed a concept plan for reuse of the airport site and ultimately a master development plan published in 1995. The plan was guided by three overarching goals:

- Create a regional job center that can contribute to the city's long-term economic health.
- Demonstrate the benefits of reducing consumption of natural resources and impacts on the natural environment.
- Provide access to social, cultural, and economic opportunities for the entire community.[171]

The plan called for a walkable, vibrant community with buildings compactly spaced on small lots to allow for large, contiguous areas of public green space. Designers met open space, stormwater management, and flood control

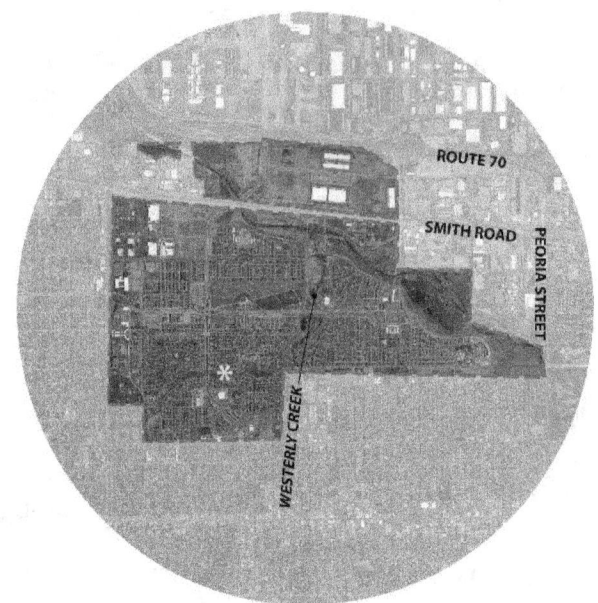

Exhibit 44. The Stapleton Airport redevelopment site encompasses 4,700 acres. The yellow asterisk marks the Greenway Park.

requirements by incorporating green infrastructure throughout the site.

C. DESIGN AND PERFORMANCE

The stormwater treatment system was designed before 2004 design regulations set by the Urban Drainage and Flood Control District in Denver went into effect. However, the system meets the 2004 requirements for water quality and flood control.[172] Design standards for the system include:

- Capture and treatment of the 80th percentile runoff event or 0.6 inches of rain.
- A 40-hour controlled drain time in detention basins.

- A runoff flow rate that matches pre- and post-development peak flows for the 2-year and 100-year storm events.

This case study focuses on the Stapleton Greenway Park between 25th Drive and East 26th Avenue, the community's first of several open green spaces that provide stormwater management for the entire development. Greenway Park's green infrastructure practices manage runoff from a 180-acre sub-watershed of Westerly Creek that is 53 percent impervious.[173] A conventional gutter and inlet collection system captures runoff from this area and conveys it to outfalls in Greenway Park. Sediment forebays at

---

[171] City and County of Denver, Stapleton Redevelopment Foundation, and Citizens Advisory Board op. cit.
[172] Denver Urban Drainage and Flood Control District. *Urban Storm Drainage Criteria Manual Volume 3 - Best Management*

Practices. 2010. http://www.semswa.org/uploads/FileLinks/a0b9436a763f4470a648b3fca2de80b3/USDCM_Volume_3.pdf.
[173] Personal communication with Dennis Arbogast, URS Corporation, on May 19, 2011.

each of the outfalls provide pretreatment. Stormwater flows from the forebays into vegetated swales that naturally convey the stormwater through a constructed wetland channel into an extended dry detention basin. The detention basin controls peak flow rates for the 100-year storm and provides controlled release to Westerly Creek through an outfall pipe that was originally part of the airport infrastructure.

This system provides flood control and can remove pollutants (including phosphorus, nitrogen, pathogens, and total suspended solids) from a large contributing drainage area without infringing on existing water rights. [174]

This community facility reduced the amount of land required for stormwater management while making the park more attractive and providing diverse soil and water conditions that can

Exhibit 45. Homes in Stapleton have views of the stormwater facility and park.

support a wide variety of vegetation, improving wildlife habitat. Other design strategies included reducing the amount of pavement and disconnecting impervious areas from the stormwater management system.

D.   COSTS AND FUNDING

Developer Forest City Stapleton, Inc. created the Park Creek Metropolitan District (PCMD) to design and construct Stapleton's infrastructure. PCMD gets funding from several sources:[175]

- Forest City pays PCMD a one-time $15,000 system development fee per acre.
- The city established a tax-increment financing district in Stapleton. Tax payments to the city taxing authorities are frozen at 2000 levels, while taxes above that base (due to rising property values) fund regional infrastructure, including parks and stormwater facilities. [176]
- The Westerly Creek Metro District levies taxes for infrastructure construction and maintenance.

- The PMCD has municipal bonding capacity and therefore has access to bond proceeds.
- Forest City provides loans to PCMD when necessary to continue the pace of infrastructure development.

The cost of installing the stormwater management system at Greenway Park to serve the 180-acre watershed totaled $420,000, including grading, outlet structures, forebays, and irrigated landscaping for the constructed stormwater wetland. [177] PCMD is responsible for maintenance, which costs approximately $1,500 per year for routine activities such as inspection and removal of trash and debris at outlet control structures. In addition, annual sediment removal costs approximately $2,000.

[174] Colorado requires the evaporative losses from a constructed permanent pool to be replaced by a similar quantity of non-surface water entitlement or water right.
[175] Roberts, Carol. "From Runways to Residences—How Stapleton is Developed." *The Front Porch*. Jan. 2014. http://fpstapleton.wpengine.netdna-cdn.com/wp-content /uploads/2014/02/How-Stapleton-is-Developed.pdf.

[176] Roberts, Carol. "How Stapleton Taxes Finance Infrastructure." *The Front Porch*. Feb. 1. 2014. http://frontporchstapleton.com/article/blighted-land-6-9-billion-development-25-years-stapleton-taxes-pay.
[177] Arbogast *op. cit.*

E. BENEFITS

Organization of the community around the greenways and parks has helped make Stapleton a very desirable area to live and work. The president and chief operating officer of Forest City Stapleton credits the emphasis on parks and community open space with helping to make Stapleton one of the best-selling master-planned communities in the United States.[178]

The vegetated swales and constructed stormwater wetland are seamlessly integrated into the park, which includes a playground, boulder-climbing area, skate park, picnic areas, and tennis courts.[179] In addition to enhancing the open space, the stormwater features create a variety of conditions that support a wide array of native vegetation important for wildlife habitat. The greenway corridors, including Greenway Park, attract small fish, frogs, whitetail deer, snowy egrets, golden eagles, killdeer, and red tail hawks.

Instead of directing stormwater directly to area waterways through underground pipes, the vegetated swales and constructed wetland filter out nutrients and sediment, improving

Jess Goff / Forest City Stapleton, Inc.

Exhibit 46. Walking and biking paths allow residents to use the open space that functions as a centralized stormwater management facility.

water quality. They also control peak flow rates, reducing erosion in receiving streams. By providing undeveloped flood plain areas with storage for 100-year flood events, the project alleviated previous flooding issues resulting from the site's limited storm sewer capacity and broad and shallow flood plain. During a major storm in 2011, the stormwater system functioned as designed, flooding the Westerly Creek channel while keeping developed areas in the community dry.[180]

F. LESSONS LEARNED

- Green infrastructure is a viable option for water quality treatment even in areas where water rights preclude certain practices. The vegetated swales and constructed wetland in this case satisfied water rights requirements not to retain, reuse, or store runoff.

- Centralized stormwater management practices can create valuable community amenities while maximizing developable land. The centralized facilities in Stapleton

are integrated into an extensive park system that provides multiple community benefits, creating value for the developer and residents.

- Having a community-driven plan in place even before soliciting developers for a large, complex infill project can reduce developers' risk by simplifying the approval process, help generate interest from prospective buyers, and ultimately facilitate implementation of the community's vision. The 1995 Stapleton

[178] Forest City Stapleton, Inc. "Stapleton Denver Named Number One Best Selling Master Planned Community in Colorado, Sixth in Nation." Press release. Jan. 27, 2015. http://www.stapletondenver.com/wp-content/uploads/2015/01/StapletonAwardRelease-2015.pdf.

[179] Chroma Design. "Stapleton Greenway Park." http://www.chromadesigninc.com/projects/greenway-1.htm. Accessed Aug. 20, 2015.
[180] Roberts, Carol. "July 7, 2011—Why Stapleton Didn't Flood." *The Front Porch.* Oct. 29, 2013. http://frontporchstapleton.com/article/july-7-2011-why-stapleton-didnt-flood-2.

Development Plan has guided the community's development for 20 years, creating certainty for all parties.

- Large, complex infill projects can require novel infrastructure financing tools.

Stapleton's developer pieced together a combination of methods to get the job done, including tax-increment financing, special assessments, developer fees, and municipal bonds.

G.   PROJECT TEAM

- **Master developer:** Forest City Stapleton, Inc.
- **Master planner:** Calthorpe Associates
- **Program manager/engineer:** URS Corporation
- **Civil engineer:** Matrix Design Group, Inc.
- **Water resources planning and design:** Matrix Design Group, Inc.

XII. MINT PLAZA
SAN FRANCISCO, CALIFORNIA

A deteriorating alley becomes a new public plaza that serves the entire community, promotes economic development, and infiltrates stormwater.

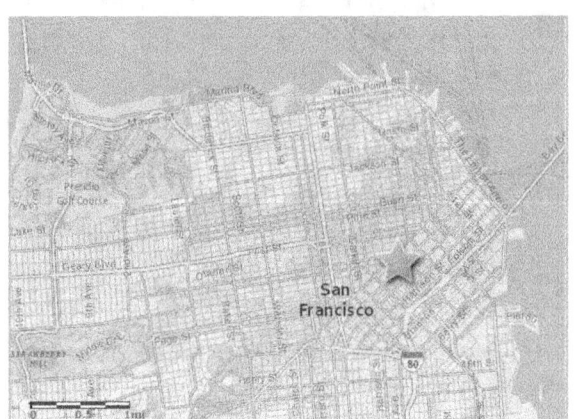

Project type:	Public plaza
Green infrastructure practices:	Rain gardens, infiltration chambers, and structural tree planters
Completion date:	2007

The Mint Plaza redevelopment project transformed a degraded alley in San Francisco into an attractive public plaza with landscape design elements that help manage stormwater. Rain gardens, infiltration basins, and structural tree planters that infiltrate stormwater keep 500,000 gallons of stormwater out of the city's combined sewer system annually. The nationally recognized[181] Mint Plaza provides the community with outdoor recreation space surrounded by restaurants and shops with offices and residences above. The plaza is frequently used for community events such as farmers markets and outdoor concerts.

The Mint Plaza project was the first of its kind in San Francisco. It involved the closing of a public street by a private developer to create a new public space. The stormwater management system employed at the site is now a prototype for how to integrate green infrastructure into highly urban sites across the city.

A. SITE CONTEXT

Mint Plaza is in San Francisco's Mid-Market neighborhood on 5th Street next to the historic U.S. Mint building (under renovation to become an event space). The Mid-Market neighborhood lies between the downtown commercial core to the east and Civic Center, an area containing many of the city's government and cultural institutions, to the northwest. In the late 1990s, the area suffered from a range of economic and social problems. In 2002, the San Francisco Planning and Urban Research Association (SPUR) created a redevelopment plan for the Mid-Market neighborhood, which aimed to revitalize the area

---

[181] Among other awards, this project received EPA's National Award for Smart Growth Achievement. See: EPA. *2010 National Award for Smart Growth Achievement Booklet.*

https://www.epa.gov/smartgrowth/2010-national-award-smart-growth-achievement-booklet.

economically, socially, and physically, building on the area's strong transit connections, cultural institutions, historic buildings, and nonprofit and social service agencies.[182]

SPUR identified the conversion of Mint Street into a combined vehicle and pedestrian space as one of several projects in the neighborhood that could help jump-start revitalization. In 2005, the San Francisco Redevelopment Agency drafted a redevelopment plan for the neighborhood that also highlighted the revitalization potential of the plaza outside of the Mint building, one of the few buildings in the neighborhood to survive the 1906 earthquake and fires.[183] Before redevelopment, this area was a neglected alley often crowded with idling tour buses and parked cars. Stormwater runoff in this area flowed directly to the city's combined sewer system.

Exhibit 47. Mint Plaza is in a highly impervious area of the city where nearly all stormwater flows to the combined sewer system.

B. PLANNING AND REGULATORY CONTEXT

The driving force behind converting the alley to a public plaza came from a local developer, Martin Building Company (MBC), which was renovating several historic warehouses adjoining the alley. MBC hoped to revitalize the area by creating a safe and welcoming outdoor space that would increase foot traffic and bring customers to local businesses. A series of public meetings to gather input into the design and generate community support for the project found that residents and other stakeholders wanted a flexible public space that could serve a variety of community needs.[184]

When the project was initiated, the city did not require redevelopment projects to manage stormwater on-site. In fact, city regulations required that stormwater from the site be conveyed to the combined sewer system. The developer and design team worked closely with the city to establish criteria for on-site water quality treatment and recharge that were then used to size the green infrastructure practices on the site.[185] City regulations also prohibited disconnection of the adjacent roof downspouts from the combined sewer system, so the designers modified their initial plans, reducing the impervious area served by green infrastructure at the site.[186] Both of these regulatory obstacles to using green infrastructure were addressed before the city issued updated stormwater design standards in 2010. MBC chose to incorporate green infrastructure into the plaza design in spite of these obstacles, recognizing that environmental sustainability would help with marketing the firm's projects and ultimately increase sales prices.[187]

[182] SPUR. *Mid-Market Street Redevelopment District: A Plan for Incremental Change.* 2002. http://www.spur.org/publications/spur-report/2002-01-16/mid-market-street-redevelopment-district.
[183] McDavid, Shelley. "Mint Plaza." In *Public Interest Design: Evaluating Public Architecture.* 2013. http://issuu.com/publicarchitecture/docs/pid_externship_report_2012-13_final/33.

[184] Gross, Jaime. "Mint Plaza in San Francisco." *Topos.* 2009: 62-64. http://www.cmgsite.com/fileadmin/cmg/home/projects/mint_plaza/CMG_Topos_67.pdf.
[185] Personal communication with Scott Cataffa, CMG Landscape Architecture, on Mar. 21, 2011.
[186] Viani, Lisa Owens. "Fresh Mint Taste." *Landscape Architecture Magazine.* Jul. 2011: 68-70.
[187] McDavid *op. cit.*

C. DESIGN AND PERFORMANCE

The site's stormwater management system includes rain gardens, subsurface infiltration chambers, and structural tree planters that manage runoff from 18,000 square feet of impervious area (Exhibit 48). Runoff flows either directly to the infiltration chambers via a brick slot drain or first to one of two rain gardens that filter out pollutants before conveying runoff to the infiltration chambers. The chambers are in the center of the plaza to avoid existing utilities, basements, and building foundations. They sit atop sandy, native soil ideal for infiltration.[188]

Six structural tree planters add green space to the plaza and additional infiltration area. The planters help keep soil loose and maintain space for a healthy root system while supporting the sidewalk above.[189] Inspection ports and cleanouts under removable pavers in the patio area and under a removable bench at the rain gardens provide access for maintenance.

Designers sized the stormwater management system to infiltrate runoff from the 5-year, 3-hour storm event, or 0.79 inches of rainfall.[190] This is the same standard the city uses to size sewer pipes, providing assurance that the green infrastructure practices could replace the conventional collection and conveyance system. During construction, the sandy soils beneath the infiltration chambers were found to have a much higher infiltration rate than the designers had expected, resulting in infiltration of up to the 25-year, 24-hour storm. Larger events drain overland to an existing gutter and inlet system in 5th Street. Over the course of a year, the new Mint Plaza removes 500,000 gallons of

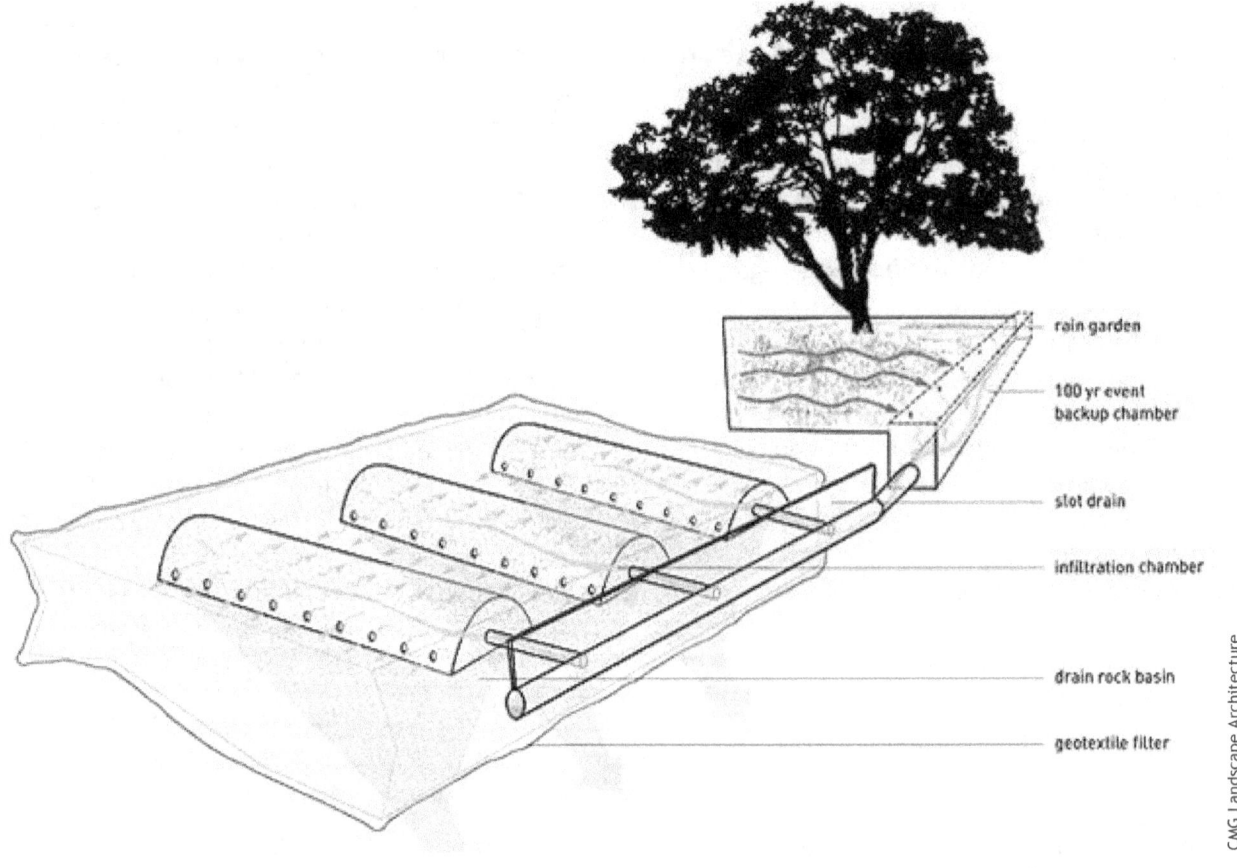

rain garden

100 yr event backup chamber

slot drain

infiltration chamber

drain rock basin

geotextile filter

CMG Landscape Architecture

Exhibit 48. Section detail of rain garden, slot drain, and subsurface infiltration chamber.

---

[188] Gross *op. cit.*
[189] Cataffa *op. cit.*

[190] Cataffa *op. cit.*

stormwater that would have otherwise gone to the city's combined sewer system.[191]

Since the construction of Mint Plaza, San Francisco passed a stormwater management ordinance[192] and issued stormwater design guidelines.[193] The standards, which apply to both new and redevelopment projects over 5,000 square feet, address the use of green infrastructure for on-site treatment and management of stormwater. Developers must design projects to capture and treat 0.75 inches of rainfall using best management practices. Mint Plaza's design exceeds this standard by capturing 0.79 inches of rainfall, and its actual performance is likely better given the unexpected high quality of the native soil.

Exhibit 49. Green infrastructure on Mint Plaza accommodates the heavy use expected from a plaza in the heart of downtown.

D.   COSTS AND FUNDING

To finance the project, MBC created a special assessment district called a Community Facilities District. California permits property owners to approve and levy a special real estate tax on their own properties to support the issuance of tax-exempt bonds by the Association of Bay Area Governments' Finance Authority for Non-profit Corporations. Proceeds from the bond sale can be used to reimburse private, for-profit developers (in this case MBC) for upfront expenses to design and build public improvements. Bonds were issued based on the increased property tax assessments levied on five surrounding privately owned properties that benefited indirectly from the development of the plaza.[194] The Community Facilities District covered the majority of the total $3.2 million cost of the plaza.

In addition, the Public Utilities Commission contributed close to $150,000, which covered a major portion of the stormwater management system;[195] a local hotel contributed $200,000 towards the plaza to meet its open space requirements;[196] and the manufacturer of the structural tree planters donated six for use at Mint Plaza as a demonstration project.[197]

In addition to creating the special tax district, MBC formed an independent, nonprofit organization, Friends of Mint Plaza, to manage maintenance and programming on the plaza, including a farmers market and arts performances.[198] The organization also hosts private, revenue-generating events at the site to pay expenses.[199]

[191] McDavid *op. cit.*
[192] City of San Francisco. *Ordinance No. 83-10.* 2010. http://www.sfbos.org/ftp/uploadedfiles/bdsupvrs/ordinances10/o0083-10.pdf.
[193] City of San Francisco. *San Francisco Stormwater Design Guidelines.* 2010. http://www.sfwater.org/Modules/Show Document.aspx?documentID=2779.

[194] Personal communication with Michael Yarne, formerly of MBC, on Mar. 30, 2011.
[195] McDavid *op. cit.*
[196] Viani *op. cit.*
[197] Cataffa *op. cit.*
[198] Gross *op. cit.*
[199] Yarne *op. cit.*

F. BENEFITS

The Mint Plaza project converted a former alley into a vibrant, public space that hosts community events for socioeconomically diverse residents in a neighborhood with limited open space. Converting the alley to a plaza created a safe public environment and contributed to revitalization of the surrounding neighborhood. New hotels, restaurants, and cafes have opened on or near the plaza, which also hosts food trucks daily.

The stormwater management system removes as much as 500,000 gallons of stormwater from the city's combined sewer system annually.[200] Given the success of the design, San Francisco has used the stormwater management system implemented at Mint Plaza as a prototype for

other projects throughout the city that integrate green infrastructure into the urban fabric in a way that benefits the environment and the neighborhood.[201]

Friends of Mint Plaza

Exhibit 50. Dance performances are some of the many activities that bring people to Mint Plaza.

G. LESSONS LEARNED

- Regulations can sometimes create obstacles to using green infrastructure. When the project was being designed, San Francisco's codes prohibited designers from directing the runoff from adjacent roofs to the plaza's infiltration chambers, reducing the project's environmental benefits. The 2010 San Francisco Stormwater Design Guidelines changed that policy to encourage developers to use green infrastructure to manage runoff on-site, ensuring that future projects will not face this limitation.

- Identifying funds for ongoing maintenance when the project was being planned was key to the public permitting process because it allayed the city's concern about who would be responsible for these costs.

- Soil testing early in the project design phase can help determine site constraints or (as in the case of Mint Plaza) identify potential cost savings that can be realized when good infiltrative soils are native to the site. Mint Plaza developers believed the project was designed to manage the 5-year storm event on-site, but thanks to the site's good soil, the plaza actually manages the 25-year storm event.

- Designing public spaces to ensure they are open and welcoming to all users can help ensure long-term community support for the project and help make revitalization more socially equitable. Public programming and movable seating independent of any of the plaza businesses have helped attract socioeconomically diverse users who do not have to patronize any businesses to use the site.[202]

---

[200] McDavid *op. cit.*
[201] Ibid.

[202] Ibid.

I.   PROJECT TEAM

- **Developer:** Martin Building Company (MBC)
- **Landscape architect:** CMG Landscape Architecture
- **Civil engineer:** Sherwood Design Engineers
- **Regulatory agencies:** City of San Francisco and San Francisco Public Utilities Commission[203]

---

[203] Ibid.

XIII. THORNTON CREEK WATER QUALITY CHANNEL
SEATTLE, WASHINGTON

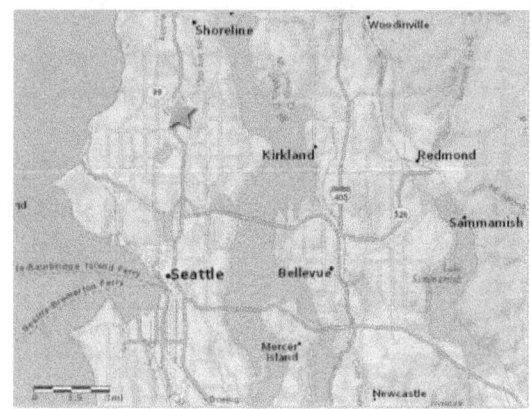

A redeveloped parking lot manages stormwater from a 680-acre sub-basin, improving water quality in a compact, densely developed area, while providing much-needed open space for residents.

Project type:	Mixed-use development
Green infrastructure practices:	Water quality channel
Completion date:	2009

This project is part of the city of Seattle's efforts to revitalize the Northgate district, provide much-needed public open space in this highly urbanized neighborhood, and improve water quality to benefit downstream habitat. The Thornton Creek water quality channel provides end-of-pipe water quality treatment for a 680-acre sub-basin of the Thornton Creek watershed. It diverts stormwater from an enclosed drainage system under the site to a series of small ponds landscaped with enhanced soils and native plants, reducing flow rates and allowing pollutants to settle out before the water reaches the creek. A stakeholder group of community, environmental, and business organizations helped define the project goals and select a design that protects the environment while allowing new development. Seattle Public Utilities built the stormwater facility as part of a joint venture with the property owners.

A.   SITE CONTEXT

The 9.1-acre project site is in the Northgate district in northeast Seattle, with the Northgate Mall and its surrounding surface parking lots to the north, a transit hub and Interstate 5 (I-5) to the west, a single- and multi-family residential neighborhood to the east, and office and commercial space to the south. Historically, the site connected surrounding wetlands with Thornton Creek's South Branch. However, in the 1950s, the proximity to I-5 transformed the area with new retail development and large surface parking lots. Part of this development included enclosing a portion of Thornton Creek between the wetlands and its headwaters in a 60-inch diameter storm pipe under 20 feet of fill and a 5-acre parking lot.[204]

The Northgate district is in the highly urban, 11.6-square-mile Thornton Creek watershed.[205] Stormwater runoff has caused erosion and water-quality problems in Thornton Creek, an important salmon habitat. The Washington Department of Ecology lists Thornton Creek as impaired for violation of bacteria, dissolved oxygen, mercury, ammonia, temperature, and pH standards.[206] Thornton Creek drains to Lake Washington in the heart of the Seattle-Bellevue metropolitan area, which also suffers from significant water quality impairment. Many in the community identified the project site's location between a highly urbanized 680-acre drainage area and the headwaters of Thornton Creek's South Branch as an ideal place to use green infrastructure to improve the quality of stormwater runoff before it reaches the creek.

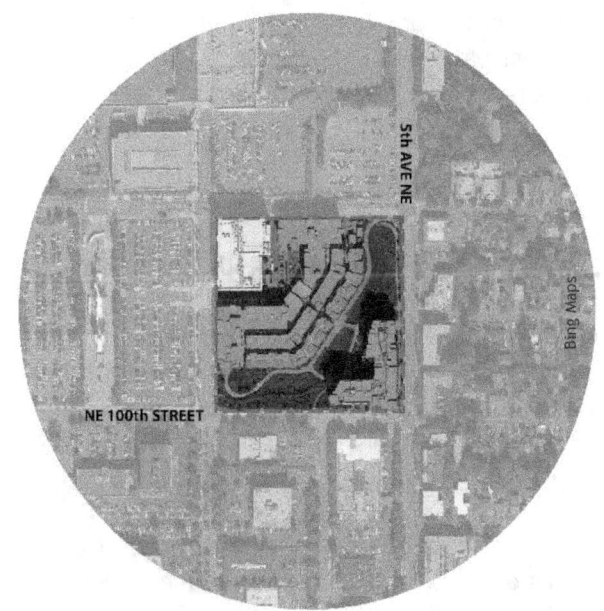

Exhibit 51. The Thornton Place development is in a highly impervious area of the city surrounded by acres of parking.

B. PLANNING AND REGULATORY CONTEXT

Seattle's 1993 *Northgate Area Comprehensive Plan* sought to change the automobile-oriented neighborhood into a mixed-use, more compactly developed area that could accommodate anticipated population growth while maintaining a high quality of life.[207] However, nearly a decade of controversy and litigation followed as property owners sought to redevelop the existing parking lot for a mix of residential and commercial uses, while environmental advocates argued for unearthing the buried creek bed.[208]

In 2003, the mayor and city council brokered an agreement that involved establishing the Northgate Stakeholder Group, with 22 members representing community, environmental, and

business interests. The city tasked the group with reaching a compromise on the fate of the site that would improve water quality in Thornton Creek, provide community open space, and generate economic development in the Northgate district.[209] The group ultimately unanimously selected a design for the site, the Thornton Creek water quality channel, that would meet all of these goals.

As part of the brokered agreement, Seattle Public Utilities entered into a joint venture with the owners of the 9-acre site to acquire 2.7 acres of the site for a stormwater management facility (the "water quality channel") positioned to maximize stormwater treatment and

[204] SvR Design. *Thornton Creek Water Quality Channel - Final Report*. Seattle Public Utilities and Restore Our Waters. 2009. http://www.seattle.gov/util/cs/groups/public/documents/webcontent/spu01_006146.pdf.
[205] Ibid.
[206] State of Washington Department of Ecology. "Water Quality Assessment and 303(d) List." http://www.ecy.wa.gov/programs/wq/303d/index.html. Accessed Jul. 23, 2015.

[207] City of Seattle. *Northgate Area Comprehensive Plan*. 1993. http://www.seattle.gov/Documents/Departments/Neighborhoods/Planning/Plan/Northgate-plan.pdf.
[208] SvR Design *op. cit.*
[209] Ibid.

development opportunities. Owners of the northwest parcel would develop it with multifamily housing and commercial space, while owners of the southeast parcel would develop it as senior housing.[210]

C.   DESIGN AND PERFORMANCE

Many green infrastructure projects distribute smaller treatment facilities throughout a watershed. However, the Thornton Creek water quality channel is designed to treat most storm flows from two sub-basins, one that covers 20 acres and one that covers 660 acres, before the runoff reaches Thornton Creek's South Branch.

Stormwater enters the facility at the upper reaches of the site from two diversion structures, one directing the majority of flows from the 660-acre sub-basin into the water quality channel, and one directing flows from the 20-acre sub-basin into the upper cascade swale, which then flows into the water quality channel. Ninety-one percent of the average annual runoff travels through the water quality channel, while peak

flows from large storm events overtop a barrier to bypass the facility and continue through the existing storm pipe. The bioswale terraces slow down the water and allow sediments and associated pollutants to settle out, while providing a beautiful landscape of native plants.[211] Space constraints led designers to modify conventional bioswale designs so that the water quality channel accepts deeper flows. It is expected to remove 40 to 80 percent of total suspended solids and associated pollutants rather than the standard 80 percent.[212] Eighty-five percent of the plants are native species. The plantings include 172 native trees, 1,792 native shrubs, and 49,000 native perennials, herbs, grasses, rushes, and sedges.[213]

Joe Mabel via Wikipedia Commons

Exhibit 52. The Thornton Creek water quality channel provides a beautiful view and a rare spot of green space next to new multifamily and senior housing.

---

[210] Ibid.
[211] Ibid.
[212] SvR Design *op. cit.*

[213] Landscape Architecture Foundation (LAF). "Thornton Creek Water Quality Channel." http://landscapeperformance .org/case-study-briefs/thornton-creek-water-quality-channel. Accessed Jul. 23, 2015.

D. COSTS AND FUNDING

Seattle Public Utilities constructed the green infrastructure facility for a total cost of $14.7 million with a low-interest loan from the Washington State Pollution Control Revolving Loan Fund (Exhibit 53).[214] Given the large impervious area treated by the water quality channel, it is a cost-effective solution for improving stream water quality and habitat, while providing a valuable public amenity.

ACTIVITY	COST
Planning	$99,026
Preliminary engineering	$166,652
Design phase (design, project management, public meetings and outreach, cost estimating, construction management)	$2,987,988
Construction	$10,738,215
Close-out	
Other agency-specific work packages (Seattle Parks and Recreation, Seattle City Light, Seattle Department of Transportation, Seattle Design Commission)	$169,744
Staff-specific work packages (real estate services, communications, grants and contracts)	$284,388
Total Project Cost	$14,446,013

Exhibit 53. Total cost estimate.

E. BENEFITS

The Thornton Creek water quality channel treats 680 acres of stormwater runoff in a highly impervious watershed, removing suspended solids and associated pollutants from 78 percent of the average annual stormwater volume. While helping to improve the water quality in Thornton Creek, it also creates additional wildlife habitat in an area that was previously a parking lot. The maintenance staff have observed many wildlife species using the channel, including stickleback fish in the sediment pools, herons, and ducks. Signs throughout the site educate visitors about the environmental benefits of the project, raising awareness of the importance of stormwater management.

Beyond these environmental benefits, the project created 2.7 acres of much-needed open space in the Northgate district and provided pedestrian links through the area, improving access to nearby transit stops. In addition, the Thornton Creek water quality channel has catalyzed as much as $200 million in private residential and commercial development.[215] The area gained 50,000 square feet of retail space and 530 condominiums and townhomes, with a mix of market-rate, subsidized, and senior housing units.[216] When Northgate's light rail station opens in 2021,[217] the area will be primed for additional transit-oriented development that will give residents and employees more commuting options. The Thornton Place homes are LEED Silver certified, and the entire site was awarded LEED for Neighborhood Development Silver certification. [218]

[214] SvR Design op. cit.
[215] LAF. "Thornton Creek Water Quality Channel." op. cit.
[216] Benfield, Kaid. "Outstanding Urbanism, Transit, & State-of-the-Art Green Infrastructure, Beautifully Mixed." Switchboard: Natural Resources Defense Council Staff Blog. Jun. 6, 2011. http://switchboard.nrdc.org/blogs/kbenfield/outstanding_urbanism_and_state.html.

[217] Dunham-Jones, Ellen. "Grey, Green, and Blue: Seattle's Northgate." AIArchitect. Nov. 8, 2013. http://www.aia.org/practicing/AIAB100516.
[218] Blanton Turner. "Thornton Place Earns LEED for Neighborhood Development Certification." Press release. Oct. 17, 2013. http://stellar.com/images/upload/_pdf_2013 1017095519_1/ThorntonPlaceLEEDNDPressRelease_101713.pdf.

F.   LESSONS LEARNED

- Green infrastructure can be a financially viable approach to treat stormwater from large drainage areas even in compact, urban settings with physically constrained sites. The project treats 78 percent of runoff from 680 acres while contributing to the vitality of the neighborhood.

- Collaboration and communication among diverse interests can lead to effective solutions for all parties even in seemingly intractable situations. Compromise and the establishment of the Northgate Stakeholders Group was critical to breaking the political logjam between advocacy groups and developers.

G.   PROJECT TEAM

- **Owner:** Seattle Public Utilities
- **Concept design:** Gaynor and Associates
- **Civil engineering and landscape architecture:** SvR Design Company
- **Hydraulic modeling and monitoring services:** Herrera Environmental

- **Structural and electrical engineering:** HDR, Inc.
- **Geotechnical engineering:** Associated Earth Sciences, Inc.[219]

---

[219] Giraldo, Greg, Masako Lo, and Melanie Davies. "Thornton Creek Water Quality Channel, Urban Water Quality and Environmental Benefits." 2nd National Low Impact Development Conference. Mar. 12-14, 2007.

www.ingramcontent.com/pod-product-compliance
Lightning Source LLC
Chambersburg PA
CBHW081303170526
45165CB00011B/3391